The ARRL 1986-1987 Advanced Class License Manual for the Radio Amateur

Edited By
Larry D. Wolfgang, WA3VIL
Bruce S. Hale, KB1MW

Contributors
Mark J. Wilson, AA2Z
Charles L. Hutchinson, K8CH
John Foss, W7KQW

Production Staff
Deborah Sandler
David Pingree
Steffie Nelson, KA1IFB
Joel Kleinman, N1BKE
Sue Fagan, Cover Design
Michelle Chrisjohn, WB1ENT
Leslie K. Bartoloth, KA1MJP

American Radio Relay League
Newington, CT 06111 USA

Foreword

Welcome to the Second Edition of *The ARRL Advanced Class License Manual for the Radio Amateur*. This manual continues in the tradition established by the First Editions of the License Manual series. The material has been carefully organized to provide a logical progression of topics, and to follow the FCC study guide for Element 4A as closely as possible. The text has been written in a manner specifically designed to teach the electronics theory and Amateur Radio techniques required to pass the Advanced-class exam. Each drawing has been carefully selected and rendered to illustrate clearly those concepts that you are likely to have trouble with as you study. Your understanding of the concepts is tested along the way by directing you to appropriate sections of the FCC Element 4A question pool. By the time you have studied the entire *Manual,* you will be familiar with the topics covered by all 507 of those questions.

The examinations for FCC licenses are now given by Volunteer Examiners, using questions taken from the Element 4A question pool, published in advance by the FCC. This procedure protects the integrity of the examination process and ensures that every applicant will receive a fair exam. You need not be afraid that the questions will take you by surprise, if you have prepared by studying and being familiar with all of the material in this book.

Chapter 1 includes the complete FCC Syllabus for the Element 4A written examination, the one you must pass to upgrade from a Technician or General class license. If you don't have a General class ticket, it also explains what other FCC elements you will need to pass. Chapter 10 lists all 507 questions from the FCC question pool released in January 1986, along with the multiple-choice answers and distractors that will be used by many Volunteer Examiner Coordinators. Between those two chapters you will find complete explanations of every electronics topic covered on the Advanced class exam. Reference material for the questions on FCC regulations is found in the ARRL publication, *The FCC Rule Book,* which contains a complete copy of Part 97, the FCC Rules for Amateur Radio, and full explanations to help you decipher the legalese. You should have a copy of that book as a companion to this one; together they provide all the information you need to pass the written test.

League publications represent a real team effort. Writing, editing, typesetting, preparing technical illustrations and final layout all require special talents. The people performing those jobs put a lot of themselves into our publications to bring you the best possible books. Within our Publications Office: the Technical Department editors and secretarial staff; the Production Department copyeditors, technical illustrators, typesetters and layout staff; and the sales and shipping staffs of our Circulation Department have all been involved in helping you obtain the proper material to study for your Advanced class license exam. We couldn't possibly list all the names on the title page, but every one of us wishes you the very best on your exam. We think you will find the new privileges to be well worth the effort. We are proud of our work, and we hope you are proud to say you are a League member.

David Sumner, K1ZZ
Executive Vice President, ARRL
Newington, Connecticut

Table of Contents

HOW TO USE THIS BOOK

To earn an Advanced class Amateur Radio license, you will have to know intermediate-level electronics theory, as well as the rules and regulations governing the Amateur Radio Service, as contained in Part 97 of the FCC Rules. You'll also have to be able to send and receive the international Morse code at a rate of 13 WPM.

This book provides a brief description of the Amateur Radio Service, and the Advanced class license in particular. The major portion of the book is designed to guide you, step by step, through the theory you must know to pass your FCC exam. The material is presented in a manner that closely follows the FCC study guide, or syllabus, printed at the end of Chapter 1. Chapter 10 contains the complete set of questions released by the FCC for use on Element 4A exams. It also includes the multiple-choice answers and distractors that will be used on exams given by many Volunteer Examiner Coordinators, including the ARRL VEC.

At the beginning of each chapter, you will find a list of "key words" that appear in that chapter, along with a simple definition for each word or phrase. As you read the text, you will find these words printed in *italic* type the first time they appear. You may want to refer back to the beginning of the chapter at that point, to learn the definition. At the end of the chapter, you may also want to go back and review those definitions.

As you study the material, you will be instructed to turn to sections of the questions in Chapter 10. Be sure to use these questions to review your understanding of the material at the suggested times. This will break the material into "bite-sized pieces," and make it easier for you to learn. Don't try to memorize all the questions. That will be impossible! Instead, by using the questions for review, you will be familiar with them when you take the test, but you will also understand the electronics theory behind the questions.

In addition to this book, you will want to purchase a copy of the ARRL publication, *The FCC Rule Book*, which covers all the rules and regulations you'll need to know. If you are just getting interested in Amateur Radio, the ARRL also offers a complete beginner's package, *Tune in the World with Ham Radio*. This package includes a cassette tape that teaches the Morse code, letter by letter. If you need to increase your code speed, ARRL also offers a complete set of cassette tapes to pick up where the *Tune in the World* tape leaves off. Even with the tapes, you'll want to tune in to the code practice sessions transmitted by W1AW, the ARRL Headquarters station. For more information about W1AW or how to order any ARRL publication or code tape, write to: ARRL Hq., 225 Main St., Newington, CT 06111.

Chapter 1

The Advanced Class License

A n Advanced class Amateur Radio license is a worthy goal to work toward. Whether you now hold a Novice, Technician or General class license, or are as yet unlicensed, you will find the extra operating privileges available to an Advanced class licensee to be worth the time spent learning about your hobby. After passing the FCC Element 4A exam, you will be able to operate on every frequency band assigned to the Amateur Radio Service. Segments of the 80, 40, 20 and 15-meter phone bands are reserved exclusively for Advanced and Extra Class operators. So if you find the General-class portions of the bands getting too crowded, just move down to these less-used segments.

To earn those extra privileges, however, you'll have to demonstrate that you know the international Morse code at a speed of 13 words per minute (WPM), intermediate electronics theory, operating practices, and the FCC's rules and regulations. This book is designed to teach you what you must know to qualify for the Advanced class Amateur Radio license.

IF YOU'RE A NEWCOMER TO AMATEUR RADIO

Earning an Amateur Radio license, at whatever level, is a special achievement. The half a million or so people in the U.S. who call themselves Amateur Radio operators, or hams, are part of a global fraternity. Radio amateurs serve the public as a voluntary, noncommercial, communication service, especially during natural disasters or other emergencies. Hams continue to make important contributions to the field of electronics. Amateur Radio experimentation is yet another reason many people become part of this self-disciplined group of trained operators, technicians and electronics experts — an asset to any country. Hams pursue their hobby purely for personal enrichment in technical and operating skills, without consideration of any type of payment.

Because radio signals do not know territorial boundaries, hams have a unique ability to enhance international goodwill. A ham becomes an ambassador of his country every time he puts his station on the air.

Amateur Radio has been around since before World War I, and hams have always been at the forefront of technology. Today hams relay signals through their own satellites in the OSCAR (Orbiting Satellite Carrying Amateur Radio) series, bounce signals off the moon, and use any number of other "exotic" communications techniques. Amateurs talk from hand-held transceivers through mountaintop repeater stations that can relay their signals to transceivers in other hams' cars or homes. Hams send their own pictures by television, talk with other hams around the world by voice or, keeping alive a distinctive traditional skill, tap out messages in Morse code. When emergencies arise, radio amateurs are on the spot to relay information to and from disaster-stricken areas that have lost normal lines of communication.

The U.S. government, through the Federal Communications Commission (FCC), grants all U.S. Amateur Radio licenses. This licensing procedure ensures operating skill and electronics know-how. Without this skill, radio operators might unknowingly cause interference to other services using the radio spectrum through improperly adjusted equipment or neglected regulations.

Who Can Be a Ham?

The FCC doesn't care how old you are or whether you're a U.S. citizen. If you pass the examination, the Commission will issue you an amateur license. Any person (except the agent of a foreign government) may take the exam and, if successful, receive an amateur license. It's important to understand that if a citizen of a foreign country receives an amateur license in this manner, he or she is a U.S. Amateur Radio operator. (This should not be confused with reciprocal licensing, which allows visitors from certain countries who hold valid amateur licenses in their homelands to operate their own stations in the U.S. without having to take an FCC exam.)

Licensing Structure

By examining Table 1-1, you'll see that there are five amateur license classes. Each class has its own requirements and privileges. The FCC requires proof of your ability to operate an amateur station properly. The required knowledge is in line with the privileges of the license you hold. Higher license classes require more knowledge — and offer greater operating privileges. So as you upgrade your license class, you must pass more challenging written examinations. The specific operating privileges for Advanced class licensees are listed in Table 1-2.

Table 1-1
Amateur Operator Licenses†

Class	Code Test	Written Examination	Privileges
Novice	5 WPM (Element 1A)	Elementary theory and regulations (Element 2)	Telegraphy in 3700-3750, 7100-7150, 21,100-21,200 and 28,100-28,200 kHz; 200 watts PEP output maximum.
Technician	5 WPM (Element 1A)	Elementary theory and regulations; general-level theory and regulations. (Elements 2 and 3)	All amateur privileges above 50.0 MHz plus Novice privileges.
General	13 WPM (Element 1B)	Elementary theory and regulations; general theory and regulations. (Elements 2 and 3)	All amateur privileges except those reserved for Advanced and Extra Class; see Table 1-2.
Advanced	13 WPM (Element 1B)	General theory and regulations, plus inter-mediate theory. (Elements 2, 3 and 4A)	All amateur privileges except those reserved for Extra Class; see Table 1-2.
Amateur Extra Class	20 WPM (Element 1C)	General theory and regulations, intermediate theory, plus special exam on advanced techniques. (Elements 2, 3, 4A and 4B)	All amateur privileges

†A licensed radio amateur will be required to pass only those elements that are not included in the examination for the amateur license currently held.

Table 1-2

Amateur Operating Privileges

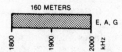

160 METERS

Amateur stations operating at 1900-2000 kHz must not cause harmful interference to the radiolocation service and are afforded no protection from radiolocation operations; see January 1986 Happenings for details.

The ARRL 160-meter band plan:
1800-1830 kHz: CW, RTTY and other narrow-band modes
1830-1840 kHz: CW, RTTY and other narrow-band modes; intercontinental QSOs only
1840-1850 kHz: CW, SSB, SSTV and other wide-band modes; intercontinental QSOs only
1850-2000 kHz: CW, SSB, SSTV and other wide-band modes

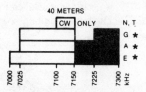

80 METERS

5167.5 kHz Alaska emergency use only. (SSB only)

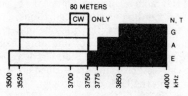

40 METERS

★ Phone operation is allowed on 7075-7100 kHz in Puerto Rico, US Virgin Islands and areas of the Caribbean south of 20 degrees north latitude; and in Hawaii and areas near ITU Region 3, including Alaska.

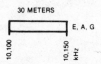

30 METERS

Maximum power limit on 30 meters is 200 watts PEP output. Amateurs must avoid interference to the fixed service outside the US.

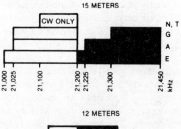

20 METERS

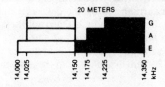

15 METERS

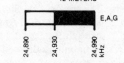

12 METERS

Amateurs must avoid interference to the fixed service outside the US.

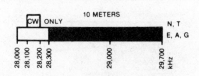

10 METERS

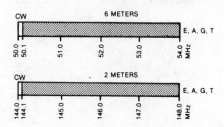

6 METERS

2 METERS

KEY

☐ = CW AND RTTY

▨ = CW, VOICE, SSTV, FAX AND RTTY

■ = CW, VOICE, SSTV AND FAX

E = EXTRA
A = ADVANCED
G = GENERAL
T = TECHNICIAN
N = NOVICE

Frequency Band
(MHz)

E,A,G,T { 220-225
420-450
902-928

The amteur bands above 928 MHz are in a state of flux at press time. For the latest information on our privileges above 928 MHz, contact ARRL HZ.

In addition, you must demonstrate an ability to receive international Morse code at 5 WPM for Novice and Technician, 13 WPM for General and Advanced and 20 WPM for Amateur Extra. It's important to stress that although you may intend to use voice rather than code, this doesn't excuse you from the code test. By international treaty, knowing the international Morse code is a basic requirement for operating on any amateur band below 30 MHz.

Learning the Morse code is a matter of practice. Instructions on learning the code, how to handle a telegraph key, and so on can be found in the ARRL *Tune in the World with Ham Radio* package. This package includes a code-learning cassette for beginners. Additional cassettes for code practice at speeds of 5, 7½, 10, 13, 15 and 20 WPM are available from the American Radio Relay League, Newington, CT 06111.

Station Call Signs

Many years ago, by international agreement, the nations of the world decided to allocate certain call-sign prefixes to each country. This means that if you hear a radio station call sign beginning with W or K, for example, you know the station is licensed by the United States. A call sign beginning with the letter U is licensed by the USSR, and so on.

International Telecommunication Union (ITU) radio regulations outline the basic principles used in forming amateur call signs. According to these regulations, an amateur call sign must be made up of one or two characters (the first one may be a numeral) as a prefix, followed by a numeral, and then a suffix of not more than three letters. The prefixes W, K, N and A are used in the United States. The continental U.S. is divided into 10 Amateur Radio call districts (sometimes called areas), numbered Ø through 9). Fig. 1-1 is a map showing the U.S. call districts.

All U.S. Amateur Radio call signs assigned by the FCC after March 1978 can be categorized into one of five groups, each corresponding to a class, or classes, of license. Call signs are issued systematically by the FCC; requests for special call signs are not granted. For further information on the FCC's call-sign assignment system,

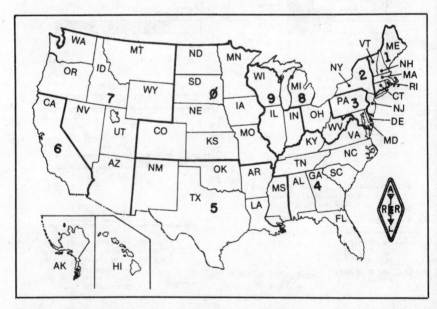

Fig. 1-1 — There are 10 U.S. call areas. Hawaii is part of the sixth call area, and Alaska is part of the seventh.

and a Table listing the blocks of call signs for each license class, see the ARRL publication, *The FCC Rule Book*. If you already have an amateur call sign, you may keep the same one when you change license class, if you wish. You must indicate that you want to receive a new call sign when you fill out an FCC Form 610 to apply for the exam or change your address.

EARNING A LICENSE

Applying for an Exam: FCC Form 610

Before you can take an FCC exam, you'll have to fill out a Form 610. This form is used as an application for a new license, an upgraded license, a renewal of a license or a modification to a license. In addition, hams who have held a valid license that has expired within the past two years may apply for reinstatement with a Form 610. Each form comes with detailed instructions. (These instructions also appear in *The FCC Rule Book*.) To obtain a Form 610, send a business-size, self-addressed, stamped envelope to: Form 610, ARRL, 225 Main St., Newington, CT 06111.

Volunteer Examiner Program

Until recently, all examinations beyond the Novice level were given by FCC examiners. A new system using amateur volunteers is now in effect. Since January 1, 1985, all U.S. amateur exams above the Novice level have been administered under this new program. *The FCC Rule Book* contains details on the volunteer examiner program. Novice exams have always been given by volunteer examiners, and that is still true. Those exams do not come under the regulations involving Volunteer-Examiner Coordinators (VECs), however.

To qualify for an Advanced class license, you must pass all elements through Element 1B and Element 4A. If you already hold a valid license, then you have credit for passing at least some of those elements, and will not have to retake them when you go for your Advanced class exam. See Table 1-1 for details.

The Element 4A exam consists of 50 questions selected by the VEC in charge of the exam session you attend. The questions are taken from a pool of 500 questions contained in FCC Bulletin PR 1035-C. Each VEC may make up their own set of possible answers, but they may not change the questions themselves in any way. The ARRL VEC, and some of the other VECs, use the multiple-choice answers printed in Chapter 10. The complete question pool from FCC Bulletin PR 1035-C, along with one correct answer and three distractors for each question, appears there. An answer key is included at the end of that chapter. The Element 4A Syllabus (outline) appears at the end of this chapter.

Finding an Exam Opportunity

To determine where and when an exam will be given, contact the ARRL Volunteer Examiner Department, or watch for announcements in the Hamfest Calendar and Coming Conventions columns in *QST*. Many local clubs sponsor exams, so they are another good source of information on exam opportunities. ARRL officials, such as Directors, Vice Directors and Section Managers receive notices about test sessions in their area. See page 8 in the latest issue of *QST* for names and addresses.

To register for an exam, send a completed Form 610 to the Volunteer Examiner team responsible for the exam session if preregistration is required. Otherwise, bring the form to the session. Registration deadlines, and the time and location of the exams, are mentioned prominently in publicity releases about upcoming sessions.

Taking The Exam

By the time examination day rolls around, you should have already prepared yourself. This means getting your schedule, supplies and mental attitude ready. Plan your schedule so you'll get to the examination site with plenty of time to spare. There's no harm in being early. In fact, you might have time to discuss hamming with another applicant, which is a great way to calm pretest nerves. Try not to discuss the material that will be on the examination, as this may make you even more nervous. By this time, it's too late to study anyway!

What supplies will you need? First, be sure you bring your current *original* Amateur Radio license, if you have one. Bring along several sharpened no. 2 pencils and two pens (blue or black ink). Be sure to have a good eraser. A pocket calculator will also come in handy. You may use a programmable calculator if that is the kind you have, but take it into your exam "empty." Don't program equations ahead of time, because you may be asked to demonstrate that there is nothing in the calculator's memory. The Volunteer Examining Team is required to check two forms of identification before you enter the test room. A photo ID of some type is best, but not required by FCC. Other acceptable forms of identification include a driver's license, a piece of mail addressed to you, a birth certificate or some other such document.

Before taking the code test, you'll be handed a piece of paper to copy the code as it's sent. The test will begin with about a minute of practice copy. Then comes the actual test; five minutes of Morse code. You are responsible for knowing the 26 letters of the alphabet, the numerals Ø through 9, the period, comma, question mark, and procedural signals $\overline{AR}$, $\overline{SK}$, $\overline{BT}$ and $\overline{DN}$. You may copy the entire text word for word, or just take notes on the content. At the end of the transmission, the examiner will hand you 10 questions about the text. Simply fill in the blanks with your answers. If you get at least 7 correct, you pass! The format of the test transmission is similar to one side of a normal on-the-air amateur conversation.

A sending test may not be required. The Commission has decided that if applicants can demonstrate receiving ability, they most likely can also send at that speed. But be prepared, just in case!

If all has gone well with the code test, you'll then take the written examination. The examiner will give all applicants a test booklet, an answer sheet and scratch paper. After that, you're on your own. The first thing to do is read the instructions. Be sure to sign your name every place it's called for. Do all of this at the beginning to get it out of the way.

Next, check the examination to see that all pages and questions are there. If not, report this to the examiner immediately. When filling in your answer sheet, make sure your answers are marked next to the numbers that correspond to each question.

Go through the entire exam, and answer the easy questions first. Next, go back to the beginning and try the harder questions. The really tough questions should be left for last. Guessing can only help, as there is no additional penalty for answering incorrectly.

If you have to guess, do it intelligently: At first glance, you may find that you can eliminate one or more "distractors." Of the remaining responses, more than one may seem correct; only one is the *best* answer, however. To the applicant who is fully prepared, incorrect distractors to each question are obvious. Nothing beats preparation!

After you've finished, check the examination thoroughly. You may have read a question wrong or goofed in your arithmetic. Don't be overconfident. There's no rush, so take your time. Think, and check your answer sheet. When you feel you've done your best and can do no more, return the test booklet, answer sheet and scratch pad to the examiner.

The Volunteer Examiner team will grade the exam right away. 74% is the passing mark. (That means no more than 13 incorrect answers.) If you are already licensed, and you pass the exam elements required to earn a higher class of license, you will receive a certificate allowing you to operate with your new privileges. The certificate has a special identifier code that must be used on the air when you use your new privileges, until your permanent license arrives from the FCC.

AND NOW, LET'S BEGIN

The section of the questions in Chapter 10 that begin with numbers 4AA- cover the rules and regulations for the Advanced class exam. You should use *The FCC Rule Book* to find the material covered by those questions. Then go over that section of the question pool to check your understanding of the rules. Perhaps you will want to study the rules a few at a time, using that as a break from your study in the rest of this book.

There you have it. The remainder of this book will provide the background in electronics theory that you will need to pass the Element 4A Advanced class written exam.

Study Guide for FCC Amateur Radio Operator License Examinations

Printed on the next page is the FCC Study Guide for the Element 4A (Advanced class) written examination. The questions in the Question Pool for Element 4A are derived from this Study Guide (or Syllabus, as it's called). As this book goes to press, the FCC has proposed a change in the Amateur Rules that may affect the syllabus. If adopted, the FCC proposal will allow Volunteer Examiner Coordinators (VECs) to maintain their own question pools for Amateur Examinations. This job has been done by the FCC until now, and the Syllabus was also released by the FCC.

If VECs are given the responsibility of question-pool maintenance, they will also be responsible for maintaining and updating their own Syllabus. The questions printed in this book are from the most recent FCC Advanced class question pool, released in January 1986.

ELEMENT 4A SYLLABUS

4A. RULES AND REGULATIONS

4AA-1) Frequency bands available to Advanced operators 97.7(a)

4AA-2) Automatic retransmission: amateur radio signals, signals from other radio services 97.3(x), 97.113, 97.126

4AA-3) Station in repeater operation 97.3(1); 97.86, 97.61(c)

4AA-4) Station in auxiliary operation 97.3(1), 97.86, 97.61(d)

4AA-5) Remote control of amateur stations 97.3(m)(2), 97.88

4AA-6) Automatic control of amateur stations 97.3(m)(3)

4AA-7) Control link 97.3(n)

4AA-8) System network diagram 97.3(u)

4AA-9) Station identification 97.84(c), (d), (e)

4AA-10) Antenna height: FAA notification, calculation of height above average terrain 97.45, 97.67(c), Appendix 5

4AA-11) Business communications 97.3(bb), 97.110

4AA-12) Remuneration for use of station 97.112

Novice operator class examinations:

4AA-13) Element 1(A) examination preparation 97.27(a)

4AA-14) Element 2 examination preparation 97.27(c), (e)

4AA-15) Examination administration 97.28(b), (c), (d)

4AA-16) Examination grading 97.29

4AA-17) Volunteer examiner requirements 97.31(a), (b)

4AA-18) Volunteer examiner conduct 97.33

4AB. OPERATING PROCEDURES

4AB-1) A3C and F3C transmission

4AB-2) Slow-scan J3F transmission

4AC. RADIO WAVE PROPAGATION

4AC-1) Sporadic-E

4AC-2) Selective fading

4AC-3) Auroral propagation

4AC-4) Radio-path horizon

4AD. AMATEUR RADIO PRACTICE

Test equipment use:

4AD-1) Frequency measurement devices

4AD-2) Dip meter

4AD-3) Performance limitations: oscilloscopes, meters, frequency counters; accuracy, frequency response, stability

4AD-4) Lissajous figures

Electromagnetic compatibility:

4AD-5) Intermodulation interference

4AD-6) Receiver desensitizing

4AD-7) Cross-modulation interference

4AD-8) Capture effect

4AE. ELECTRICAL PRINCIPLES

Concepts:

4AE-1) Reactive power

4AE-2) Series and parallel resonance

4AE-3) Skin effect

4AE-4) Fields, energy storage, electrostatic, electromagnetic

Mathematical relationships:

4AE-5) Resonant frequency, bandwidth, and Q of R-L-C circuits

4AE-6) Phase angle between voltage and current

4AE-7) Power factor, phase angle

4AE-8) Effective radiated power, system gains and losses

4AE-9) Replacement of voltage source and resistive voltage divider with equivalent circuit consisting of a voltage source and one resistor (An application of Thevenin's Theorem, used to predict the current supplied by a voltage divider to a known load.)

4AF. CIRCUIT COMPONENTS

Physical appearance, types, characteristics, applications, and schematic symbols

4AF-1) Diodes; Zener, tunnel, varactor, hot-carrier, junction, point contact, PIN

4AF-2) Transistors; NPN, PNP, junction, unijunction, power, germanium, silicon

4AF-3) Silicon controlled rectifier, triac

4AF-4) Light-emitting diode, neon lamp

4AF-5) Crystal-lattice SSB filters

4AG. PRACTICAL CIRCUITS

4AG-1) Voltage-regulator circuits; discrete and integrated

4AG-2) Amplifiers: Class A, AB, B, C; characteristics

4AG-3) Impedance-matching-networks; Pi, L, Pi-L

4AG-4) Filters; constant K, M-derived, band-stop, notch, modern-network-theory, Pi-section, T-section, L-section

4AG-5) Oscillators; types, applications, stability

4AG-6) Frequency synthesizer

Transmitter and receiver circuits — purpose; function:

4AG-7) Modulators: AM, FM, balanced

4AG-8) Transmitter final amplifiers

4AG-9) Detectors, mixer stages

4AG-10) RF and IF amplifier stages

4AG-11) Mechanical filters

Calculation of voltages, currents, and power in common amateur radio circuits:

4AG-12) Common emitter class A transistor amplifier; bias network, signal gain, input and output impedances

4AG-13) Common collector class A transistor amplifier; bias network, signal gain, input and output impedances

Circuit design selection of component values:

4AG-14) Voltage regulator with pass transistor and Zener diode to produce given output voltage

4AG-15) Select coil and capacitor to resonate at given frequency

4AH. SIGNALS AND EMISSIONS

4AH-1) Emission types A3C, C3F, A3F, F3C, J3F

4AH-2) Modulation methods

4AH-3) Deviation ratio

4AH-4) Modulation index

4AH-5) Electromagnetic radiation

4AH-6) Wave polarization

4AH-7) Sine, square, sawtooth waveforms

4AH-8) Root-mean-square value

4AH-9) Peak-envelope-power relative to average

4AH-10) Signal-to-noise ratio

4AI. ANTENNAS AND FEEDLINES

4AI-1) Antenna gain, beamwidth

4AI-2) Trap antennas

4AI-3) Parasitic elements

4AI-4) Radiation resistance

4AI-5) Driven elements

4AI-6) Antenna efficiency

4AI-7) Folded, multiple-wire dipoles

4AI-8) Velocity factor

4AI-9) Electrical length of a transmission line

4AI-10) Voltage and current nodes

4AI-11) Mobile antennas

4AI-12) Loading coil; base, center, top

KEY WORDS

Amplitude modulation — A method of placing some information on an electrical signal. The amplitude of the signal is varied in a way that is controlled by the information signal. For FAX (A3C) or SSTV (J3F or A3F) transmissions it refers to a method of superimposing picture information on the radio-frequency signal.

Bandwidth — The frequency range over which a signal is stronger than some specified amount below the peak signal level. For example, a certain signal is at least half as strong as the peak power level over a range of ± 3 kHz, so it has a 3-dB bandwidth of 6 kHz.

Crystal oscillator — An oscillator in which the main frequency-determining element is the mechanical resonance of a piezoelectric crystal (usually quartz).

Facsimile (FAX) — The process of scanning pictures or images and converting the information into signals that can be used to form a likeness of the copy in another location.

Fast-scan TV (ATV) — A television system used by Amateurs that employs the same video-signal standards as commercial TV.

Flying-spot scanner (FSS) — A device that uses a moving light spot to scan a page. The intensity of the reflected light is sensed by a photoelectric cell, generating a signal that contains information about what is on the page.

Frequency modulation — A method of placing some information on an electrical signal. The instantaneous frequency of a sine-wave carrier is varied by an amount that depends on the frequency of the information signal at that instant. For FAX (F3C) or SSTV (J3F) transmissions it refers to a method of superimposing picture information on the radio-frequency carrier. G3C and G3F refer to a method of varying the phase of the carrier wave. There is no practical difference in the way you receive phase-modulated signals from frequency-modulated ones.

Gray scale — A photographic term that defines a series of neutral densities (based on the percentage of incident light that is reflected from a surface), ranging from white to black.

Horizontal synchronization pulse — Part of a TV signal used by the receiver to keep the CRT electron-beam scan in step with the camera scanning beam. This pulse is transmitted at the beginning of each horizontal scan line.

Negative-format modulation — A FAX system that uses a minimum modulation level to produce white and a maximum modulation level to produce black.

Photocell — A solid-state device in which the voltage and current-conducting characteristics change as the amount of light striking the cell changes.

Photodetector — A device that produces an amplified signal that changes with the amount of light striking a light-sensitive surface.

Phototransistor — A bipolar transistor constructed so the base-emitter junction is exposed to incident light. When light strikes this surface, current is generated at the junction, and this current is then amplified by transistor action.

Positive-format modulation — A FAX system that uses a maximum modulation level to produce white and a minimum modulation level to produce black.

Slow-scan television (SSTV) — A television system used by Amateurs to transmit pictures within a signal bandwidth allowed on the HF bands by the FCC. It takes approximately 8 seconds to send a single frame with SSTV.

SSTV scan converter — A device that uses digital signal-processing techniques to change the output from a normal TV camera into an SSTV signal or to change a received SSTV signal to one that can be displayed on a normal TV.

Sync — Having two or more signals in step with each other, or occurring at the same time. A pulse on a TV or FAX signal that ensures the transmitted and received images start at the same point.

Synchronous motor — An ac electric motor whose rotation speed is controlled by the frequency of the electric energy used to power it.

Tuning-fork oscillator — An oscillator in which the main frequency-determining element is the mechanical resonance of a tuning fork.

Vertical synchronization pulse — Part of a TV signal used by the receiver to keep the CRT electron-beam scan in step with the camera scanning beam. This pulse returns the beam to the top edge of the screen at the proper time.

Chapter 2

Operating Procedures

As an Advanced class Amateur Radio licensee, you will be expected to know standard operating practices used on the amateur bands. For the Novice class license (FCC Element 2), you had to learn about a few procedures, such as the RST reporting system, how to tune a transmitter and how to zero beat a received signal. To pass the Technician/General class exam (Element 3) you were required to learn more standard operating procedures, such as how to operate radiotelephone, how to use RTTY, and repeater techniques. These represented some of the new operating privileges that go with the higher-class license. As you increase your Amateur Radio knowledge and experience while working toward the Advanced class license, you will want to try some of the more exotic modes available to you.

Questions on the Advanced class exam will cover image communications using *facsimile (FAX)* and *slow-scan television (SSTV)* techniques. This chapter deals with the standard operating practices used by amateurs on these modes.

FACSIMILE SYSTEMS

Facsimile (FAX) is the earliest operating system used for image transmission over radio. The pictures have high resolution, which is achieved by using from 800 to several thousand lines per frame. To keep the transmitted-signal *bandwidth* within the narrow limits allowed by the FCC on the HF bands, relatively long transmission times are used for FAX pictures. Depending on the system in use, it may take from a bit more than three minutes up to 15 minutes to receive a single frame.

Many amateurs use FAX to receive weather-satellite photos that are transmitted direct from space. News services also use this mode to distribute ''wirephotos'' throughout the world, and FAX systems are becoming increasingly common in business and law-enforcement offices to transfer documents, signatures, fingerprints and photographs.

Transmission

A typical FAX transmitter consists of a small-diameter drum driven by a *synchronous motor.* The material to be transmitted is wrapped around the drum, which rotates at a constant speed. This speed is carefully controlled, usually by a *crystal* or *tuning-fork oscillator,* which generates an ac signal that can be amplified to run the motor. Drum speeds are expressed in revolutions/min, which is equivalent to the scanning line frequency (lines/min). A small spot of light is focused on the printed material, and reflected light is picked up by a *photocell, photodetector* or *phototransistor.* The light source and sensor are attached to a carriage that moves along the drum to provide a scan of the entire picture area.

Voltage variations from the light pickup are amplified and used to modulate an audio subcarrier signal. Either *amplitude modulation (A3C)* or *frequency modu-*

lation (F3C or G3C) methods may be used, with FM being the standard for amateur HF operation and some of the press services using AM. *Positive-format modulation* means that modulation is proportional to brightness. Black produces little or no modulation (up to about 4%), and white produces 90 to 100% modulation. For *negative-format modulation* this is reversed, with white producing minimum modulation and black producing a maximum amount.

There are three important characteristics for any FAX system: drum-rotation speed, carriage or paper feed rate (which determines the scan density of the picture) and modulation characteristics. These characteristics must be the same for the receiving and transmitting stations. Table 2-1 lists standards used by some of the common FAX services. The 240 lines/min drum speed is a good rate for amateur HF use because a detailed picture (800 lines) can be transmitted in about 3.3 minutes. If a drum speed of 120 lines/min is used, then it takes approximately 6 minutes to transmit a complete picture. Scan density is usually expressed as a number of lines per inch. This parameter determines the aspect ratio of the picture (image width/height) and the transmission time for one frame.

FAX operation is permitted by the FCC in the voice segments of all amateur bands above 3.775 MHz. Note that there is no voice operation allowed on 30 meters, though. The FCC rules require you to identify your station every 10 minutes during a transmission or series of transmissions, and at the beginning and end of the communications. This ID must be either voice or CW. Sending a FAX picture that includes your call sign is not sufficient. Amateurs using certain digital codes or *fast-scan TV (ATV)* are permitted to identify in that mode, but this does not apply to FAX or SSTV operation.

Reception

There are several methods available for displaying a received FAX picture. One system uses a drum, with a single sheet of paper for each picture. See Fig. 2-1 for

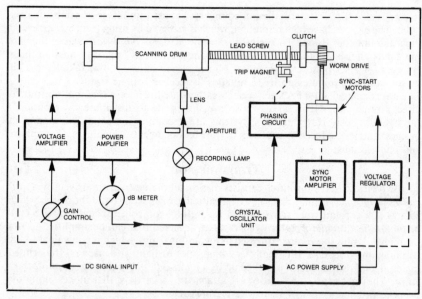

Fig. 2-1 — Block diagram of a photographic-paper-type of facsimile receiving unit. Other types of receiving units are described in the text.

Table 2-1

Standards for Various FAX Services

Service	Drum Speed (r/min or lines/min)	Size (in)	Scan Density (lines/in)
WEFAX Satellite	240	11	75
APT Satellite	240	11	166
Weather Charts	120	19	96
Wire photos	90	11	96
Wire photos	180	11	166

Fig. 2-2 — Example of a FAX picture received from a Japanese amateur by Jerry Grokowsky, WA9HCZ. The motor speed on Jerry's FAX receiver was controlled by a crystal-oscillator circuit designed to produce 120-Hz ac to run the motor.

a block diagram of one type of receiver. This system operates in a manner similar to the FAX transmitter. Another type of receiver is the continuous-feed recorder, which uses paper on a roll to provide a continuous chart type of printout.

FAX recorders may use a variety of paper types, also. One that uses photographic paper or film requires a darkroom to process the received image. Others use electrostatic paper in which a white coating is burned off a black base layer. By varying the amount of coating removed, the recorder can reproduce a good *gray scale* (shades ranging from black to white). These pictures can approach photographic quality. If a current is passed through special chemically treated paper, a dark line can be "burned" on the paper. This method can also produce a reasonable gray scale. Yet another system uses a type of stylus and a carbon-paper arrangement to draw the picture. The image quality with this system is marginal, and not recommended.

A few adventurous amateurs are using computer systems to receive FAX pictures. High-resolution TV-monitor displays provide a 16-level gray scale and good resolution, and permanent hard copy is possible with a graphics printer. The computer makes it possible to display inverted images, perform various types of image enhancement, and even produce "color" images.

To receive FAX pictures, the characteristics of your system must match those of the transmitting station. Amateurs commonly use 240 line/min or 120 line/min systems. Even if you are sure that you are using the same scan rates as the transmitting station, you may have trouble receiving pictures. The motor speeds of the two stations must be synchronized, so it is helpful to have a means to vary the motor-speed control frequency a bit. Fig. 2-2 shows a FAX "QSL" from Japan, received by Jerry Grokowsky, WA9HCZ.

Equipment

Obtaining surplus commercial FAX equipment is an excellent way to get on this mode, since the equipment is well constructed and inexpensive. Ideally, you should find a machine that has multiple drum speeds, variable scanning densities and switchable modulation methods. Be sure to obtain a complete set of technical manuals, however. The most difficult aspect of putting a two-way amateur FAX station together is finding a FAX transmitter. It is possible to fabricate a simple drum system with a light source, photo transistor, op-amp buffer and SSTV modulator. As an alternative, you may want to investigate putting together a computer system to both transmit and receive FAX pictures.

[Before proceeding, study FCC examination questions with numbers that begin 4AB-1. Review this section as needed.]

SLOW-SCAN TELEVISION TRANSMISSION

A commercial TV signal has a bandwidth greater than 5 MHz. Since this is more frequency space than is available in all of the amateur bands below 6 meters combined, it is obvious that if we want to send TV pictures over the air, we cannot use commercial-style techniques. Another stumbling block is the FCC's requirement that we limit our image signals to no more than the bandwidth of a single-sideband voice signal in certain portions of the HF bands.

Slow-scan TV (SSTV) is a form of television that has a very slow scan rate. This makes it possible to keep the bandwidth within the required limitations. A regular fast-scan TV signal produces 30 frames every second, but SSTV takes 8 seconds to send one frame! Amateurs can only use fast-scan TV (ATV) on frequencies above 420 MHz. If you think of ATV as watching home movies by radio, then SSTV is like a slide show over the air. Because of the slow scan rate, SSTV can be used in the Advanced and Extra Class portions of any amateur band above 3.5 MHz. Many DX stations are now equipped for this mode, and some amateurs have contacted more than 100 countries on SSTV.

Technical Details

The bandwidth of any radio signal is directly proportional to the data rate. The more information that is being transmitted in a given amount of time, the higher the bandwidth. This is true regardless of the signal type: CW, voice or other. Instead of the normal rate of 30 frames per second and 525 lines per frame, an SSTV picture takes 8 seconds for one frame and has only 120 scan lines. This works out to 15 picture lines per second instead of 15,750 lines per second. It is easy to see that this slower data rate results in a greatly reduced bandwidth requirement.

The video information is normally transmitted as a frequency-modulated audio subcarrier. This would be designated as an F3F, J3F or G3F emission, although A3F is also permitted. A 1500-Hz signal is used to produce black, and a 2300-Hz signal produces white on the TV screen. Frequencies between these two represent shades of gray. *Horizontal* and *vertical synchronization pulses* must be added to keep transmitting and receiving systems in sync. These pulses are sent as bursts of 1200-Hz tones. Since this represents a shade "blacker than black," the sync pulses do not show up on the screen. Table 2-2 summarizes these SSTV standards. The horizontal sync pulse is included in the time to send one line, but the vertical sync pulse adds 30 ms of "overhead." So it takes a bit more than 8 seconds to actually transmit a picture frame.

Equipment

An SSTV camera produces a variable-frequency audio tone. Higher tones represent light areas, lower tones represent dark areas. To transmit the SSTV picture, simply feed this signal into the microphone input of your SSB transmitter. To receive a picture, just tune in a signal on an SSB receiver and feed the audio signal to your SSTV monitor.

Table 2-2

SSTV Standards

Frame time	8 seconds
Lines per frame	120
Time to send one line	67 ms
Duration of horizontal sync pulse	5 ms
Duration of vertical sync pulse	30 ms
Horizontal and vertical sync frequency	1200 Hz
Black frequency	1500 Hz
White frequency	2300 Hz

Even if you don't have a camera to start with, you can become involved in SSTV operation. Since the transmitted pictures are just audio tones, they can be recorded off the air using a normal audio tape recorder. You can play the signals back through your monitor for viewing again later and you can feed the audio into your microphone jack to retransmit the pictures. Another SSTVer may be willing to prerecord some pictures from his camera for your use.

If you want to be able to create some pictures of your own, but still don't want to invest in a camera, you might consider a *flying-spot scanner (FSS)*. This device transmits pictures by scanning a slide transparency with a spot of light. A detector senses how much light goes through the slide, and this is converted to an SSTV signal. You can build up a library of slides for transmission, any of which can be selected almost instantly. Another advantage of having an FSS is that it frees the camera for other things, and eliminates the need to readjust the lighting and focus as you change from subject to subject.

With the popularity of home computers, there are some programs available to generate SSTV audio tones, so this is another way to get in on the SSTV fun. Computer graphics open up many new possibilities for creating your own SSTV pictures.

One very important equipment consideration is that SSTV is a 100%-duty-cycle transmission mode. This means your transmitter will be producing full power for the full picture-transmission time. Most SSB transmitters and amplifiers will have to be run at reduced power output to avoid overheating the components.

Color SSTV

Early experiments with color transmission involved the use of red, green and blue filters between the camera and the image and between the monitor screen and a color Polaroid® camera. A set of three frames of the same picture would be sent in succession, with the filter colors switched each time. The receiving station would record each new frame on the same piece of film, and if the camera alignment was not disturbed, a color picture would result. While this system works, it is not very reliable or elegant.

With the advent of scan converters to change a fast-scan image to slow-scan format, new methods became feasible. Either a black-and-white camera with filters or a color camera can be used to produce the picture. The scan converter includes a memory area for each color, so three frames are transmitted and recombined for display on a color monitor at the receiving end.

Fig. 2-3 is a block diagram of such a system. Various formats are being used. You can send a complete frame of each color, or a single line can be repeated for the red, green and blue information, then the next line sent and so on. If only one memory is used with the black-and-white camera, then you would have a basic *SSTV scan-converter* system.

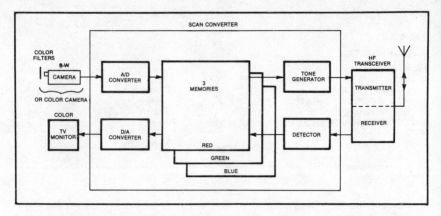

Fig. 2-3 — Block diagram of one type of color SSTV system. There are several formats for transmitting the red, green and blue picture information. A similar system with only one memory could be used just for black-and-white pictures.

Camera Procedures

It takes some time to focus a slow-scan camera because you can't see the effects of readjusting the focus right away. Having a high-contrast vertical line in the picture helps. As the picture is scanned from top to bottom, you can keep your eye on the vertical line on your monitor as you adjust the camera focus. A fast-scan-camera signal can be converted to the slow-scan format. With this type of system, you can watch the picture on a fast-scan screen while you adjust the focus, then switch to the slow-scan format for transmission.

Lighting is critical with SSTV because the reproducible brightness range is rather limited. You should adjust the contrast and brightness controls on your monitor by using a test pattern. You might do this by asking another station to transmit a test pattern over the air for you, or by having such a pattern recorded on tape. Test patterns typically have several vertical bars of various shades of gray. Proper monitor calibration is achieved when the darkest bar is just dark enough to blacken the screen while the brightest bar just reaches maximum brightness.

Now you are ready to adjust the brightness and contrast controls on your camera. Set them so the darkest part of the image is black on your calibrated monitor screen and the brightest areas just reach maximum brightness.

Your subjects don't all have to be either live or on tape. Some SSTVers find it convenient to mount the camera in front of an easel. Various photographs, drawings or lettered signs can be placed on the easel for convenient display. The variety of subjects that you may choose to transmit is almost endless, but you will find that high-contrast black-and-white pictures work best. They should not be cluttered with too much fine detail. Some possibilities are a close-in shot of you at your operating position, cartoons (commercial or homemade) and a couple of frames with your call and perhaps your name and QTH printed on them.

General Operating

SSTV has good potential for providing service to the general public by means of third-party traffic. For example, the scientists working on the Antarctic ice pack don't see their families for months — except by amateur slow-scan TV. People can write or call almost anywhere in the world today using commercial services. But how often can grandma and grandpa see their grandchildren from 3000 miles away? SSTV provides this opportunity, and without a charge!

SSTV is legal in the Advanced and Extra Class voice segments of the 75, 40, 20 and 15-meter bands, and on all voice bands above 28 MHz. Standard calling frequencies are 3.845, 7.171, 14.230, 21.340 and 28.680 MHz. The most popular bands are 20 and 75 meters.

It is customary to send two or three frames of each subject to ensure that the other station receives at least one good-quality picture. Many operators send a couple of frames followed by voice comments concerning the picture, alternating between picture and voice signals on one frequency. Some of the more elaborate stations are equipped to transmit video and audio simultaneously, with the voice on one sideband and the video on the other. These independent sideband (ISB) signals can be copied with two receivers or a single receiver with two IF filters, two detectors and two audio amplifiers.

An SSTV signal must be tuned in properly so that the picture will come out with the proper brightness and so the 1200-Hz sync pulses are detected properly. If the signal is not "in sync," the picture will appear wildly skewed. The easiest way to tune a signal is to wait for the operator to say something and then fine tune him or her. With some experience, you may find that you are able to zero in on an SSTV signal by listening to the sync pulses and watching for proper synchronization on the screen. Many SSTV monitors are equipped with some type of tuning aid.

If you want to record slow-scan pictures off the air, there are two ways of doing it. One is to tape record the audio signal for later playback. The other method is to take a picture of the image on the SSTV screen using a camera. A close-up lens will be helpful if you choose this method. A Polaroid instant camera lets you see the results almost immediately. A light-tight hood to fit between the monitor screen and the camera allows you to take pictures without darkening the room.

In general, SSTV operating procedures are quite similar to those used on SSB. The FCC requires you to identify your station at the beginning and end of a transmission or series of transmissions and at not more than a 10-minute interval during a single long transmission. You may not use SSTV pictures to identify your station in meeting this requirement. CW or voice ID is required. This also makes it easier for non-SSTV stations to tell who you are. You may spark their interest in the mode when they hear you talking about the pictures you are sending.

[Before moving on to Chapter 3, study those FCC examination questions with numbers beginning 4AB-2. Review this section as needed.]

Key Words

Absorption — The loss of energy from an electromagnetic wave as it travels through any material. The energy may be converted to heat or other forms. Absorption usually refers to energy lost as the wave travels through the ionosphere.

Aurora — A disturbance of the atmosphere around the earth's poles, caused by an interaction between electrically charged particles from the sun and the earth's magnetic field. Often a display of colored lights is produced, which is visible to those who are close enough to the magnetic-polar regions. Auroras can disrupt HF radio communication and enhance VHF communication. They are classified as visible auroras and radio auroras.

Equinoxes — One of two spots on the earth's orbital path around the sun at which it crosses a horizontal plane extending through the center of the sun. The vernal equinox marks the beginning of spring and the autumnal equinox marks the beginning of autumn.

K index — A geomagnetic-field measurement taken at Boulder, Colorado, and updated every few hours. The K index can be used to indicate HF propagation conditions.

Multipath — A fading effect caused by the transmitted signal traveling to the receiving station over more than one path.

Polarization — A property of an electromagnetic wave that tells whether the electric field of the wave is oriented vertically or horizontally. The polarization sense can change from vertical to horizontal under some conditions, and can even be gradually rotating either in a clockwise (right-hand-circular polarization) or a counterclockwise (left-hand-circular polarization) direction.

Radio horizon — The position at which a direct wave radiated from an antenna becomes tangent to the earth's surface. Note that as the wave continues past the horizon, the wave gets higher and higher above the surface.

Selective fading — A variation of radio-wave field intensity that is different over small frequency changes. It may be caused by changes in the material that the wave is traveling through or changes in transmission path, among other things.

Solar wind — Electrically charged particles emitted by the sun, and traveling through space. The wind strength depends on how severe the disturbance on the sun was. These charged particles may have a sudden impact on radio communications.

Sporadic-E propagation — A type of radio-wave propagation that occurs when dense patches of ionization form in the E layer of the ionosphere. These "clouds" reflect radio waves, extending the possible VHF communications range.

Summer solstice — One of two spots on the earth's orbital path around the sun at which it reaches a point farthest from a horizontal plane extending through the center of the sun. With the north pole inclined toward the sun, it marks the beginning of summer in the northern hemisphere.

Tropospheric ducting — A type of radio-wave propagation whereby the VHF communications range is greatly extended. Certain weather conditions cause portions of the troposphere to act like a duct or waveguide for the radio signals.

Wind shear — An effect in which high-velocity winds flowing in opposite directions in the upper atmosphere cause discontinuities in the stratosphere. This results in densely ionized patches in the E layer of the ionosphere.

Winter solstice — One of two spots on the earth's orbital path around the sun at which it reaches a point farthest from a horizontal plane extending through the center of the sun. With the north pole inclined away from the sun, it marks the beginning of winter in the northern hemisphere.

Chapter 3

Radio-Wave Propagation

A s you advance in your knowledge of Amateur Radio, and study for a higher license class, you earn the privilege of using additional frequencies and operating modes. You will also learn more about how radio waves travel from one place to another to carry the information you are transmitting. To pass FCC Element 3 for a Technician or General class license, you learned many basic properties of radio-wave propagation. To pass the FCC Element 4A exam for an Advanced class license, you must study some types of propagation that are a little more difficult to understand. But once you do, you will be ready to put that knowledge to use and enhance your enjoyment of our hobby.

The material in this chapter will prepare you for the questions on section 4AC of the FCC syllabus. This material covers four areas of propagation: sporadic-E, selective fading, auroral propagation and radio-path horizon. Questions 4AC-1.1 through 4AC-4.5 in the question pool test the information presented here. You will be directed to those questions in Chapter 10 at appropriate places throughout this chapter. See *The ARRL Handbook* and *The ARRL Antenna Book* for further reading on propagation.

SPORADIC-E

Sporadic-E propagation, also known as E-skip or E_s, occurs when radio waves are reflected by dense patches of ionization that form in the E region of the ionosphere, approximately 50 miles above the earth. These ionized patches, or clouds as they are often called, form randomly and usually last up to a few hours at a time. Their transient nature, and the E-layer altitude, account for the name sporadic-E. Sporadic-E is, however, of a different origin and has different communications potential than E-layer propagation that affects mainly the 1.8- through 14-MHz bands.

Cloud Formation

Although the causes of sporadic-E clouds are not completely understood, it is generally believed that *wind shear* is involved. High velocity winds, flowing in opposite directions at slightly different altitudes, produce shears. A shear, in this case, is an area where the two winds are in contact. The tearing forces of the wind produce sharp and violent "rips" in the continuity of the stratosphere. The effect of these strong winds flowing in opposite directions seems to be to tear electrons off the gas molecules that make up the air. As a result, strong charges of static electricity develop around these shears. The effect is similar to producing static electricity by rubbing your shoes rapidly across a carpet in dry winter weather. In the presence of the geomagnetic field, ambient ionization around these shears is redistributed and compressed into a thin, dense layer. Data from rockets penetrating E_s regions confirm the electron density, wind velocity and height parameters predicted by this model.

Another related theory states that violent thunderstorm activity is involved.

Thunderstorms are accompanied by strong electric charges and high-velocity winds that may cause ionization in the atmosphere to be redistributed and compressed into dense clouds.

E_s is directly associated with terrestrial and meteorological phenomena, but there is no apparent relationship to sunspot activity. This mode of propagation is equally good, regardless of whether the sunspot cycle is at maximum or minimum.

E_s Opening

Sporadic-E is most prevalent in the equatorial regions, but it is common in the temperate latitudes as well. Although it can occur at any time, sporadic-E propagation happens most often in the months around the *summer* and *winter solstices*. North American amateurs look forward to two E_s seasons each year — May through August and December through January. The long and short seasons are reversed in the southern hemisphere. Fig. 3-1 shows the relative amount of sporadic-E activity during the year in northern latitudes.

The duration and extent of E_s openings tends to be greater in the summer season. Late June and early July is usually the peak period. There are E_s openings almost every day then, and country-wide openings may last for many hours at a time. E_s propagation is most common during the midmorning and early evening hours, but it may extend around the clock at times.

The MUF of intense E_s clouds is a function of their ionization density. The upper frequency limit for sporadic-E propagation is not known. It is observed fairly often up to about 100 MHz and has been reported on frequencies as high as 200 MHz. E_s is most common on the 28- and 50-MHz amateur bands. Although sporadic-E is most often associated with 50-MHz work because it is the main DX mode there, it is often the only means of long-distance 28-MHz propagation during low-sunspot years, when F-layer propagation is not consistently available. E-skip is not nearly as common on 144 MHz; according to some estimates, only 5 percent of the 6-meter openings are accompanied by 2-meter openings. On this band, openings are brief and extremely variable. Two-meter operators take full advantage of those occasions, however; thousands of 2-meter E_s contacts are made each summer.

Layer height and electron density determine the skip distance. On 28 and 50 MHz, sporadic-E propagation is most common over distances of 400 to about 1300 miles. This is known as single-hop E-skip. Sometimes, E_s clouds form simultaneously in two or more locations, making multiple-hop (commonly called double-hop) E-skip possible. Double-hop E_s may extend the working range to as much as 2600 miles.

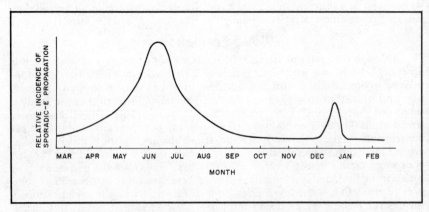

Fig. 3-1 — There are two times during the year that sporadic-E propagation is most likely.

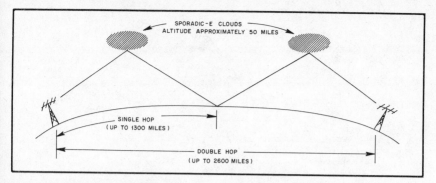

Fig. 3-2 — Single-hop sporadic-E propagation affords communications possibilities up to 1300 miles. Double-hop sporadic-E extends the range up to 2600 miles.

Signals are usually heard from intermediate distances at such times. See Fig. 3-2.

Ionization develops rapidly, with effects first showing on lower frequencies. Amateurs interested in 6-meter sporadic-E work often monitor the lower VHF television channels and the 10-meter amateur band for signs of activity. They also monitor the FM broadcast band and TV channels 5 and 6 for indications of 2-meter sporadic-E propagation.

As ionization density increases and the MUF rises, the skip distance on a given band shortens. When the E-skip distance on 10-meters shortens to 500 miles or less, it is likely that E_s will be noted on 6 meters. Likewise, very short skip on 6 meters may signal a 2-meter opening.

On the air, signals arriving by sporadic-E are often extremely strong. Signal strength varies as the E_s clouds shift, and at times signals may go from S-meter-pinning strength to barely audible in a matter of seconds. Also, the areas that are workable are very specific, changing as the clouds move. For example, a station in New England may work only stations in eastern Iowa, with excellent signal reports in both directions and then not hear them at all a few minutes later as the propagation shifts to stations in Minnesota. During intense E_s openings, minimal power is required for reliable communications. Often stations using a few watts and a dipole will work nearly the same range as stations running the legal power limit and a large Yagi array.

[Study FCC examination questions with numbers that begin 4AC-1. Review this section as needed.]

FADING

Fading is a general term used to describe variations in the strength of a received signal. It may be caused by natural phenomena such as constantly changing ionospheric-layer heights, variations in the amount of *absorption,* or random *polarization* shifts when the signal is refracted. Fading may also be caused by man-made phenomena such as reflections from passing aircraft and ionospheric disturbances caused by exhaust from large rocket engines.

Multipath

A common cause of fading is an effect known as *multipath.* Serveral components of the same transmitted signal may arrive at the receiving antenna from different directions, and the phase relationships between the signals may cancel or reinforce each other. This effect is illustrated in Fig. 3-3. Multipath fading is responsible for

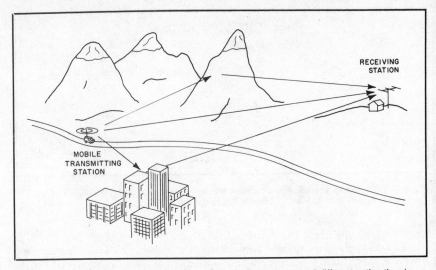

Fig. 3-3 — If a signal travels from a transmitter to a receiver over several different paths, the signals may arrive at the receiver slightly out of phase. The out-of-phase signals alternately cancel and reinforce each other, and the result is a fading signal. This effect is known as multipath fading.

the effect known as "picket fencing" in VHF communications when a signal from a mobile station will have a rapid fluttering quality. This fluttering is caused by the change in the paths taken by the transmitted signal to reach the receiving station as the mobile station moves.

Multipath effects can occur whenever the transmitted signal follows more than one path to the receiving station. Some examples of this with HF propagation would be if part of the signal goes through the ionosphere and part follows a ground-wave path, or if the signal is split in the ionosphere and travels through different layers before reaching the receiving station. It is even possible to experience multipath fading if part of the signal follows the long path around the earth to reach the receiver, while part of the signal follows the direct short-path route. When the transmitted signal reaches the receiver over several paths, the end result is a variable-strength signal.

Selective Fading

Selective fading is a type of fading that occurs when the wave path from a transmitting station to a receiving station varies with very small changes in frequency. It is possible for components of the same signal that are only a few kilohertz apart (such as the carrier and the sidebands in an AM signal) to be acted upon differently by the ionosphere, causing modulation sidebands to arrive at the receiver out of phase. The result is distortion that may range from mild to severe.

Wideband signals, such as high-quality FM and double-sideband AM, suffer the most from selective fading. The sidebands may have different fading rates from each other or from the carrier. Distortion from selective fading is especially bad when the carrier of an FM or AM signal fades while the sidebands do not. In general, the distortion from selective fading is worse with FM than it is with AM. SSB and CW signals, which have a narrower bandwidth, are affected least by selective fading.

[Now study FCC examination questions with numbers that begin 4AC-2. Review this section as needed.]

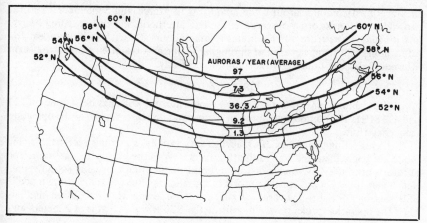

Fig. 3-4 — The possibility of auroral propagation decreases as distance from the geomagnetic North Pole increases.

AURORAL PROPAGATION

Auroral propagation occurs when VHF radio waves are reflected from ionization associated with an auroral curtain. It is a VHF and UHF propagation mode that allows contacts up to 1300 miles. Auroral propagation occurs for stations near the northern and southern polar regions, but the discussion here is limited to auroral propagation in the northern hemisphere.

Aurora is caused by a large-scale interaction between the earth's magnetic field and electrically charged particles ejected from the sun during disturbances on the sun's surface. These particles form a *solar wind,* which drifts through space. If this solar wind travels in a direction toward earth, then the magnetic field around the earth will pull the charged particles toward the magnetic poles. A visible aurora, often called the northern lights, is caused by the collision of these solar-wind particles with oxygen and nitrogen molecules in the upper atmosphere.

When the oxygen and nitrogen molecules are struck by the solar-wind particles, they are ionized. When the electrons that were knocked loose recombine with the molecules, light is produced. The extent of the ionization determines how bright the aurora will appear. At times, the ionization is so strong that it is able to reflect radio signals with frequencies above about 20 MHz. This ionization occurs at an altitude of about 70 miles, very near the E layer of the ionosphere. Not all auroral activity is intense enough to reflect radio signals, so a distinction is made between a visible aurora and a radio aurora.

The number of auroras (both visible and radio) varies with geomagnetic latitude. Generally, auroral propagation is available only to stations in the northern states, but, on occasion, extremely intense auroras reflect signals from stations as far south as the Carolinas. Auroral propagation is most common for stations in the northeastern states and adjacent areas of Canada, which are closest to the earth's magnetic pole. This mode is rare below about 32° north latitude in the southeast and about 38° to 40° N in the southwest. See Fig. 3-4.

The number and distribution of auroras is related to the solar cycle. Auroras occur most often during sunspot peaks, but the peak of the auroral cycle appears to lag the solar-cycle peak by about two years. Intense auroras can, however, occur at any point in the solar cycle.

Auroras also follow seasonal patterns. Although they may occur at any time, they are most common around the *equinoxes* in March and September. Auroral propagation is most often observed in the late afternoon and early evening hours, and it usually lasts from a few minutes to many hours. Often, it will disappear for a few hours and reappear around midnight. Major auroras often start in the early afternoon and last until early morning the next day.

Using Aurora

Most common on 10, 6 and 2 meters, some auroral work has been done on 220 and 432 MHz. The number and duration of openings decreases rapidly as the operating frequency rises.

The reflecting properties of an aurora vary rapidly, so signals received via this mode are badly distorted by multipath effects. CW is the most effective mode for auroral work. The CW tone is distorted, and is most often a buzzing sound rather than a pure tone. For this reason, auroral propagation is often called the "buzz mode." SSB is usable for 6-meter auroral work if signals are strong; voices are often intelligible if the operator speaks slowly and distinctly. SSB is rarely usable at 2 meters and above.

All stations point their antennas north during the aurora, and, in effect, "bounce" their signals off the auroral zone. The optimum antenna heading varies with the position of the aurora and may change rapidly, just as the visible aurora does. Constant probing with the antenna is recommended to peak signals, especially if the beamwidth is narrow. Usually, an eastern station will work the greatest distance to the west by aiming as far west of north as possible. The opposite applies for western stations working east. This does not always follow, however, so you should keep your antenna moving to find the best heading. See Fig. 3-5.

Developing auroral conditions may be observed by monitoring signals in the region between the broadcast band and 5 MHz or so. If, for example, signals in the 75-meter band begin to waver suddenly (flutter or sound "watery") in the afternoon

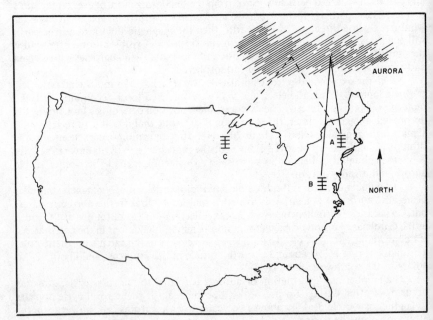

Fig. 3-5 — To work the aurora, stations point their antennas north. Station A may have to beam west of north to work station C.

or early evening hours, a radio aurora may be beginning. Since auroras are associated with solar disturbances, you can often predict one by listening to WWV's Geoalert broadcasts at 18 minutes after each hour. In particular, the *K index* may be used to indicate auroral activity. K-index values of 3, and rising, indicate that conditions associated with auroral propagation are present in the Boulder, Colorado area. Timing and severity may be different elsewhere, however. Maximum occurrence of radio aurora is for K-index values of 7 to 9. See Figs. 3-6 and 3-7.

On 6 and 2 meters, the buzzing sound that is characteristic of signals reflected by the aurora may be heard even on local signals when both beams are aimed north. The range of stations workable via aurora extends from local out to 1300 miles, but distances of a few hundred miles are most common. Range depends to some extent on transmitter power, antenna gain and receiver sensitivity, but as with most modes of operation, patience and operator skill are important.

[At this point, you should proceed to Chapter 10 and study those FCC examination questions that begin with numbers 4AC-3. Review this section as needed.]

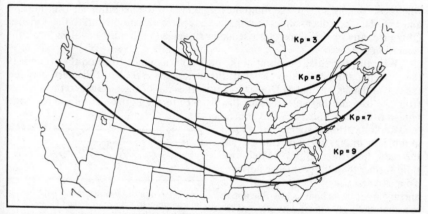

Fig. 3-6 — As the intensity of auroral activity increases, stations farther south are able to take advantage of it. The K_p numbers refer to the K Index. See Fig. 3-7.

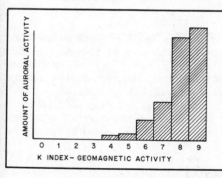

Fig. 3-7 — It is possible to predict auroral activity by monitoring the K Index during WWV broadcasts at 18 minutes past the hour. A K Index of 3 or greater, and rising, may indicate auroral activity. K Indices of 8 and 9 produce the most auroral activity.

RADIO PATH HORIZON

In the early days of VHF amateur communications, it was generally believed that space-wave communications depended on direct line-of-sight paths between the communicating station antennas. However, after experimentation was done with good equipment and antennas, it became clear that radio waves are bent or scattered in several ways, and that reliable VHF and UHF communications are possible with stations beyond the visual horizon. The farthest point to which space waves will travel directly is called the *radio horizon*.

Under normal conditions, the structure of the atmosphere near the earth causes radio waves to bend into a curved path that keeps them nearer to the earth than true straight-line travel would. This is why the distance to the radio horizon exceeds the distance to the visual, or geometric, horizon. The distance to the radio horizon can be approximated by assuming that the waves travel in straight lines, but that the earth's radius is increased by one third. On this assumption, the distance from the transmitting antenna to the horizon is given by:

$$D(mi) = 1.415 \times \sqrt{H(ft)} \qquad \text{(Eq. 3-1)}$$

or

$$D(km) = 4.124 \times \sqrt{H(m)} \qquad \text{(Eq. 3-2)}$$

where D is the distance to the radio horizon and H is the height of the transmitting antenna. The formula assumes that the earth is perfectly smooth out to the horizon. Of course, any obstructions that rise along any given path must be taken into consideration.

With this formula, the point at the horizon is assumed to be on the ground. If the receiving antenna is also elevated, the maximum space-wave distance between the two antennas is equal to $D + D1$; that is, the sum of the distance to the horizon from the transmitting antenna plus the distance to the horizon from the receiving antenna. Fig. 3-8 illustrates this principle.

Radio-horizon distances are shown graphically in Fig. 3-9. You can see that the radio horizon is approximately one third farther than the geometric horizon. To make best use of the space wave, the antenna must be as high as possible above the surroundings. This is why stations located high and in the clear on hills or mountaintops have a substantial advantage on the VHF and UHF bands, compared with stations in lower areas.

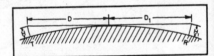

Fig. 3-8 — The distance, D, to the radio horizon from an antenna of height H is given by the formulas in the text. The maximum distance over which two stations may communicate by space wave is equal to the sum of their distances to the horizon.

The space wave goes essentially in a straight line between the transmitter and the receiver, so antennas that are low-angle radiators (that is, antennas that concentrate the energy toward the horizon) are best. Energy radiated at angles above the horizon will pass over the receiving antenna. The polarization of both the receiving and transmitting antennas should be the same because the polarization of a space wave remains constant as it travels.

As a radio wave travels in space, it collides with air molecules and other particles. When it collides with these particles, the radio wave gives up some kinetic energy. This is why there is a limit to distances that may be covered by space-wave communications.

Normal VHF propagation is usually limited to distances of approximately 500 miles. This is the normal limit for stations using high-gain antennas, high power and sensitive receivers. At times, however, VHF communications are possible with stations up to 2000 or more miles away. Certain weather conditions cause ducts in the troposphere, simulating propagation within a waveguide. Such ducts cause VHF radio waves to follow the earth's curvature for hundreds, or thousands, of miles. This form of propagation is called *tropospheric ducting*.

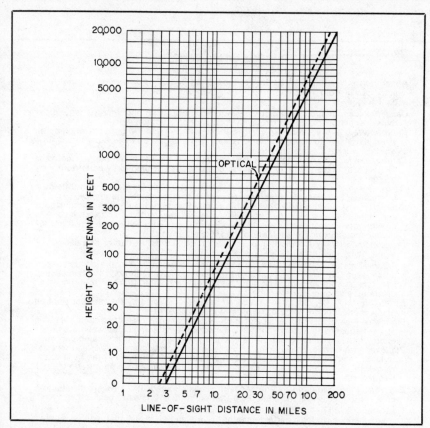

Fig. 3-9 — Distance to the radio horizon from an antenna of given height is indicated by the solid line. The broken line indicates the distance to the visual, or geometric, horizon.

The possibility of propagating radio waves by tropospheric ducting increases with frequency. Ducting is rare on 50 MHz, fairly common on 144 MHz and more common on higher frequencies. Gulf Coast states see it often, and the Atlantic Seaboard, Great Lakes and Mississippi Valley areas see it occasionally, usually in September and October.

[Before going on to Chapter 4, turn to Chapter 10 and study FCC examination questions with numbers that begin 4AC-4. Review this section as needed.]

Key Words

Absorption wavemeter — A device for measuring frequency or wavelength that takes some power from the circuit under test when the meter is tuned to the same resonant frequency.

Capacitive coupling (of a dip meter) — A method of transferring energy from a dip-meter oscillator to a tuned circuit by means of an electric field.

Capture effect — An effect especially noticed with FM and PM systems whereby the strongest signal to reach the demodulator is the one to be received. You cannot tell whether weaker signals are present.

Cathode-ray tube — An electron-beam tube in which the beam can be focused on a luminescent screen. The spot position can be varied to produce a pattern on the screen.

Cavity — A high-Q tuned circuit that passes energy at one frequency with little or no attenuation but presents a high impedance to another nearby frequency.

Circulator — A passive device with three or more ports or input/output terminals. It can be used to combine the output from several transmitters to one antenna. A circulator acts as a one-way valve to allow radio waves to travel in one direction (to the antenna) but not in another (to the transmitter).

Cross modulation — A type of intermodulation caused by the carrier of a desired signal being modulated by an unwanted signal.

D'Arsonval meter movement — A type of meter movement in which a coil is suspended between the poles of a permanent magnet. Dc flowing through the coil causes it to rotate an amount proportional to the current. A pointer attached to the coil indicates the amount of deflection on a scale.

Desensitization — A reduction in receiver sensitivity caused by the recevier front end being overloaded by noise or RF from a local transmitter.

Dip meter — A tunable RF oscillator that supplies energy to another circuit resonant at the frequency that the oscillator is tuned to. A meter indicates when the most energy is being coupled out of the circuit by showing a dip in indicated current.

Duplexer — A device, usually employing cavities, to allow a transmitter and receiver to be connected simultaneously to one antenna. Most often, as in the case of a repeater, the transmitter and receiver operate at the same time on different frequencies.

Electromagnetic radiation — Another term for electromagnetic waves, consisting of an electric field and a magnetic field that are at right angles to each other.

Frequency counter — A digital-electronic device that counts the cycles of an electromagnetic wave for a certain amount of time and gives a digital readout of the frequency.

Inductive coupling (of a dip meter) — A method of transferring energy from a dip-meter oscillator to a tuned circuit by means of a magnetic field between two coils.

Intermodulation distortion (IMD) — A type of interference that results from the mixing of integer multiples of signals in a nonlinear stage or device. Mixing products, which can interfere with desired signals on the mixed frequencies, result.

Isolator — A passive attenuator in which the loss in one direction is much greater than the loss in the other.

Lissajous figure — An oscilloscope pattern obtained by connecting one sine wave to the vertical amplifier and another sine wave to the horizontal amplifier. The two signals must be harmonically related to produce a stable pattern.

Marker generator — A crystal oscillator designed to produce a signal rich in harmonics, usually for the purpose of calibrating a receiver dial.

Oscilloscope — A device using a cathode-ray tube to display the waveform of an electric signal with respect to time or as compared with another signal.

White noise — A random noise that covers a wide frequency range across the RF spectrum. It is characterized by a hissing sound in your receiver speaker.

Zero beat — The condition that occurs when two signals are at exactly the same frequency. The beat frequency between the two signals is zero.

Chapter 4

Amateur Radio Practice

N o Amateur Radio operator will be able to do much operating without using some test equipment occasionally. Elaborate equipment is required only for the most advanced techniques; most operators will not need such equipment for many years after they begin operating. The first section of this chapter describes some commonly used test equipment. Following this description are sections on a few commonly encountered problems and their solutions.

TEST EQUIPMENT

This section describes the use of marker generators (including frequency standards), frequency counters, dip meters and oscilloscopes. Included are discussions on the performance limitations of these test-equipment items. Uses of other test equipment, such as multimeters, vacuum-tube voltmeters and FET multimeters are covered in the *Technician/General Class License Manual*.

Marker Generator

A *marker generator* is sometimes referred to as a crystal calibrator. In Amateur Radio, it is used almost exclusively to locate band edges.

A marker generator emits a signal, usually a harmonic, at every 25, 50, or 100-kHz point through the LF, MF and HF bands, and often well into the VHF and UHF bands. These harmonic signals are used to verify that the frequency readout on the receiver or transceiver is correct or, if incorrect, what adjustments must be made. Most marker generators use a 100-kHz frequency standard (quartz crystal) and a trimmer capacitor to effect minor changes in the actual marker frequency. The quartz-crystal frequency standard is an accurate and stable reference. If 50- or 25-kHz marker signals are required, a frequency-divider circuit is employed. See Fig. 4-1.

Several nations, including the United States, broadcast frequency standards 24 hours per day. In the U.S., these standards are broadcast from WWV in Fort Collins, Colorado, and from WWVH in Kekahu (Kauai), Hawaii. Thus, the first step in the use of a marker generator is to *zero beat* its signal with that of WWV or WWVH, usually on 5, 10 or 15 MHz. The trimmer capacitor is used to zero-beat the signals. Then the marker generator can be used to check where the band edges appear on the radio dial.

A marker generator might be built into a transceiver or receiver, or it might be used as a separate external accessory. If it is used with a transceiver, you must be sure the receiver incremental tuning (RIT) control is set at zero before marking band edges. If a separate receiver and transmitter are in use, it is a good idea to monitor

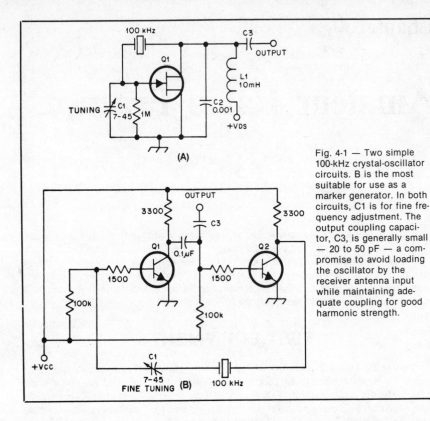

Fig. 4-1 — Two simple 100-kHz crystal-oscillator circuits. B is the most suitable for use as a marker generator. In both circuits, C1 is for fine frequency adjustment. The output coupling capacitor, C3, is generally small — 20 to 50 pF — a compromise to avoid loading the oscillator by the receiver antenna input while maintaining adequate coupling for good harmonic strength.

the transmitted signal on a calibrated receiver (preferably while using a dummy antenna), to be sure the transmitter frequency readout is correct. In almost all equipment, it is the carrier frequency that is displayed, whether the carrier is transmitted or suppressed. So, be certain that all transmitted sidebands, regardless of emission mode, are within the band.

Many manufacturers of frequency standards specify the maximum percentage error that can be expected. You should assume that at least that much error will be present. It is good practice to operate far enough from the band edge that, even in the worst possible case, the sidebands still will be inside the band.

Frequency Counter

A *frequency counter*, once considered almost a luxury item, is now an integral part of most commercially made amateur transceivers. The name is completely descriptive: A frequency counter counts the number of cycles per second (hertz) of a signal, and displays that number on a digital readout. Some frequency counters even incorporate voice synthesizers that announce the frequency on command.

Modern transceivers often have a frequency counter built in to measure and display the operating frequency, although not all digital readouts employ a frequency counter. If a counter is used, then the frequency readout will be as accurate as the counter itself. Some rigs sample some information from the control circuitry and then calculate the operating frequency for display. In such a case, if some part of the radio circuitry is not working properly, you may get an erroneous display.

You can also use a frequency counter to measure signal frequencies throughout a piece of equipment, and to make fine adjustments to tuned ciricuits. In this case, a frequency counter becomes a valuable piece of test equipment.

Although usually quite accurate, a frequency counter should be checked regularly against WWV, WWVH or some other frequency-standard broadcast. Frequency counters that operate well into the gigahertz range are available. Counters that operate at VHF or UHF, and sometimes those that operate near the top of the HF range, usually employ a prescaler. The prescaler uses logic circuitry to divide the frequency prior to counting, greatly extending the useful range of the frequency counter. A typical counter is illustrated in block-diagram form in Fig. 4-2.

The accuracy of frequency counters is often expressed in parts per million (ppm). Even after checking the counter against WWV, you must take this possible error into account. The readout error can be as much as:

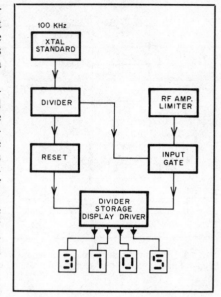

Fig. 4-2 — Frequency-counter block diagram.

$$Error = f(Hz) \times \frac{counter\ error}{1,000,000} \qquad \text{(Eq. 4-1)}$$

With a counter error of five parts per million at 30 MHz, the counter might err by as much as 150 Hz:

$$Error = 30,000,000\ Hz \times \frac{5}{1,000,000} = 150\ Hz$$

Therefore, the operator must be sure that the sidebands are at least 150 Hz inside the band edge.

The limiting factor for the accuracy of a frequency counter is the crystal reference oscillator, or time base, built into the circuit. The more accurate this crystal is, the more accurate the readings will be. Close-tolerance crystals are used, and there is usually a trimmer capacitor across the crystal so the frequency can be set exactly once it is in the circuit.

[Study the FCC examination questions in Chapter 10 with numbers that begin 4AD-1. Review this section as needed.]

Dip Meter

Once called a grid-dip meter because it employed a vacuum tube with a meter to indicate grid current, this handy device is actually an RF oscillator. Modern solid-state *dip meters* use FETs. The principle of operation is that when the meter is brought near a circuit resonant at the meter oscillation frequency, that circuit will take some power from the dip meter. That will result in a slight drop in the meter reading. Fig. 4-3 is the schematic diagram of a simple dip meter.

The most frequent amateur use of a dip meter is to determine the resonant frequency of an antenna or antenna traps. Dip meters can also be used to determine

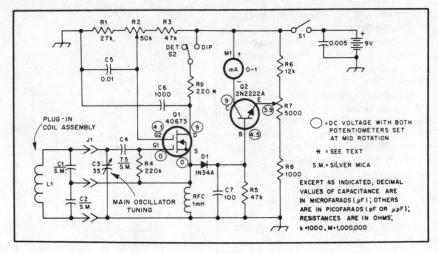

Fig. 4-3 — Schematic diagram of a dual-gate MOSFET dip meter. L1, C1 and C2 make up plug-in tuned circuits to change the operating frequency.

the resonant frequency of other circuits. The circuit under test should have no signal or power applied to it while you couple the dip meter to the circuit. Sometimes you simply need a small signal at some specific frequency to inject into a circuit. The tunable RF oscillator of a dip meter is ideal for this task.

Some dip meters are equipped with a switch to disable the power supply and insert a diode into the circuit. Then the meter can be used as an *absorption wavemeter*. In such a case, of course, resonant frequency would be indicated by a slight increase (rather than a decrease) in the meter reading. The circuit under test must have normal supply voltage and signals applied to it to use an absorption wavemeter. Be careful when you work around live circuits.

Although dip meters are relatively easy to use, the touchiest part of their use is coupling the oscillator coil to the circuit being tested. This coupling should be as loose as possible and still provide a definite, but small, dip in the current when coupled to a circuit resonant at the dip-meter-oscillator frequency. Coupling that is too loose will not give a dip sufficient to be a positive indication of resonance.

Whenever two circuits are coupled, however loosely, each circuit affects the other to some extent. If the coupling is loose, the effect will be small and will not create a significant change in the resonant frequency of either circuit. Too tight a coupling, however, almost certainly will yield a false reading on the dip meter.

Dip meters are usually coupled to a circuit by allowing the oscillator-coil field to cut through a coil in the circuit under test. This is called *inductive coupling*. The energy is transferred through the magnetic fields of the coils. Sometimes it is not possible to couple the meter to an inductor, so *capacitive coupling* is used. In this case, the dip-meter coil is simply brought close to an element in the circuit, and the capacitance between the components couples a signal to the circuit. This means that it is the electric fields between the components that transfers the energy. Fig. 4-4 indicates the methods of coupling a dip meter to a circuit.

The procedure for using a dip meter is to bring the dip-meter coil within a few inches of the circuit to be tested and then sweep the oscillator through the frequency band until the meter needle indicates a dip. This dip should be symmetrical — that is, the needle should move downward and upward at about the same speed when the oscillator is tuned through resonance. A jumpy needle indicates that the coupling

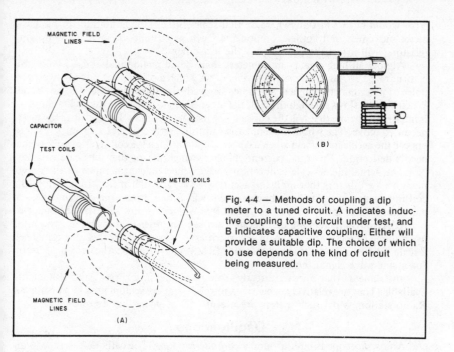

MAGNETIC FIELD LINES

CAPACITOR

TEST COILS

DIP METER COILS

MAGNETIC FIELD LINES

(A)

(B)

Fig. 4-4 — Methods of coupling a dip meter to a tuned circuit. A indicates inductive coupling to the circuit under test, and B indicates capacitive coupling. Either will provide a suitable dip. The choice of which to use depends on the kind of circuit being measured.

is too tight, there is a problem with the dip meter itself or in the circuit being tested, or there is a strong influence from yet another circuit active in the vicinity of the test area. The test should be repeated several times, to be sure the dip occurs at the same point each time and thus accurately indicates the resonance point.

If no dip appears during this check, try another coil on the dip meter. The actual resonant frequency of the circuit being tested might be far from that expected — too far even to be within the resonant-frequency range of the coil originally used on the dip meter. Some experience with dip meters usually is required before you can be sure you are not getting a false reading, caused by coupling that is too tight or by another resonant circuit near or connected to the one you are testing.

If you want to check the resonant frequency of an antenna, a noise bridge and receiver can also be used to indicate resonance. This setup will also give an indication of antenna impedance. If a noise bridge is available, it may be better to learn how to use it for checking your antenna resonant frequency.

Another problem with dip meters is the possibility of reading a harmonic rather than the fundamental frequency. The dip for a harmonic, of course, will not be as deep as that for the fundamental. Nevertheless, it is a good idea to take a reading at double, triple, one-half and one-third the apparent resonant frequency. By doing so, you can be sure the original dip actually occurs at the true resonant frequency.

[Go to the FCC examination questions with numbers that begin 4AD-2, and study them before proceeding. Review this section as needed.]

Performance Limitation of Meters

The accuracy of most meters is specified as a percentage of full scale. If the specification states that the meter accuracy is within two percent of full scale, the possible error anywhere on a scale of 0 to 10 V is two percent of 10 V, or 0.2 V. The actual value read anywhere on that scale could be as much as 0.2 V above or below the indicated meter reading. If you are using a 0- to 100-mA scale, the possible error would be ±2 mA anywhere on that scale. Besides the limitations of the basic meter movement, any other components, such as current- or voltage-multiplying resistors, will affect the accuracy of the instrument.

Almost all high-quality multimeters that use an analog dial and a needle also incorporate a *D'Arsonval meter movement*, in which a coil is positioned between the poles of a permanent magnet. An indicator needle is attached to this moveable coil. When current flows through the coil, the magnetic field will cause the coil to move. It is important to note that it is the current flowing through the coil that causes the meter needle to move. The whole movement is carefully balanced, and a spring is used to oppose the needle movement when a current flows through the coil. This provides equal needle deflections for equal current changes, which means that the scale is linear.

The actual meter-scale calibration and the mechanical tolerance of the movement are two factors that limit the accuracy of a particular meter. In addition, the coil impedance controls how much current will flow through the movement when a certain voltage is applied across it, and that will affect how sensitive the meter is.

D'Arsonval meters are dc-operated devices. In order to measure ac, you must use a diode to rectify the ac. A bridge rectifier is often used for this purpose. Then a scale can be calibrated to read effective (RMS) voltage. Such a scale is only useful for a sine-wave signal, however.

Instruments that operate directly from ac to measure current and voltage are available. They are relatively expensive, however, and are seldom used in an Amateur Radio station. Most such meters are useful only at power-line frequencies.

Oscilloscope

An *oscilloscope* is built around a *cathode-ray tube*. The cathode-ray tube in an oscilloscope differs greatly from that in a television receiver. In fact, about the only thing the tubes have in common is that both use an electron beam focused on a fluorescent screen. The oscilloscope electron beam is controlled by electrostatic charges on the vertical and horizontal deflection plates.

When a voltage is applied to the plates, the beam is pulled toward the plate with the positive charge and repelled by the one with the negative charge. (Remember that electrons are negatively charged ions.) With two pairs of plates at right angles, the beam can be moved from side to side and up and down. Frequencies far into the RF region can be applied to the plates, with the use of very little power.

In the absence of deflection voltages, the oscilloscope controls are adjusted so a small bright spot appears in the center of the screen. Then, an ac voltage applied to the horizontal plates causes the spot to move from side to side (Fig. 4-5). Usually, the time-base signal or horizontal sweep voltage is applied to these plates, causing the spot to move toward the right at a steady speed. The sweep-voltage frequency (a voltage with a sawtooth waveform) is selected by the user. The signal to be analyzed is applied to the vertical deflection plates. The speed at which the spot moves in any direction is exactly proportional to the rate at which the voltage is changing.

In the course of one ac cycle, the spot will move upward while the voltage applied to the vertical plates is increasing from zero. When the positive peak of the cycle is reached, and the voltage begins to decrease, the spot will reverse its direction and move downward. It will continue

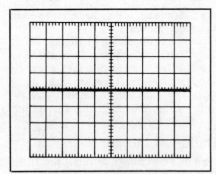

Fig. 4-5 — If the voltage applied to the vertical plates of the oscilloscope is zero, only the horizontal line created by the sweep oscillator will be visible.

in this direction until the negative peak is reached. Then, as the voltage increases toward zero, the spot moves upward again. If the horizontal sweep voltage is moving the spot horizontally at a uniform rate of speed at the same time, it is easy to analyze the signal waveform applied to the vertical deflection plates. See Fig. 4-6.

There are many uses for an oscilloscope in an amateur station. This instrument is often used to display the output waveform of a transmittter during a two-tone test. Such a test can help you determine if the amplifier stages in your rig are operating in a linear manner. An oscilloscope can also be used to display signal waveforms during troubleshooting procedures.

Another frequent use of the oscilloscope in amateur practice is the comparison of two signals, one of which is applied to the vertical deflection plates, and the other to the horizontal plates. This procedure produces a *Lissajous figure* on the scope. When sinusoidal ac voltages are applied to both sets of deflection plates, the resulting pattern depends on the relative amplitudes, frequencies and phases of the two voltages. If the ratio between the two frequencies is constant and can be expressed as two integers, the Lissajous pattern will be stationary.

Examples of some simple Lissajous patterns are shown in Fig. 4-7. The frequency ratio is found by counting the number of loops along two adajacent edges. Thus, in the pattern shown at C, there are three loops along a horizontal edge and only one along the vertical, so the ratio of the vertical frequency to the horizontal frequency is 3:1. In part E, there are four loops along the horizontal edge and three along the vertical edge, producing a ratio of 4:3. Assuming that

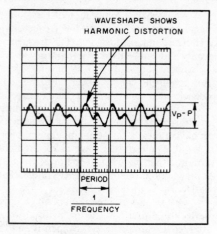

Fig. 4-6 — A typical pattern resulting when a complex waveform is applied to the vertical plates, and the sweep-oscillator output is applied to the horizontal plates.

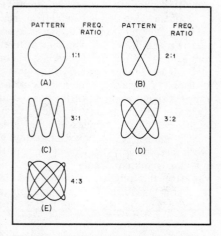

Fig. 4-7 — Lissajous figures and corresponding frequency ratios for a 90° phase relationship between the voltages applied to the two sets of deflection plates.

a known frequency is applied to the horizontal plates and an unknown frequency is applied to the vertical plates, the relationship is given by:

$$f_2 = \frac{n_2}{n_1} f_1 \qquad \text{(Eq. 4-2)}$$

where

f_1 is the known frequency
f_2 is the unknown frequency
n_1 is the number of loops along a vertical edge
n_2 is the number of loops along a horizontal edge.

An important application of Lissajous figures is the calibration of audio-frequency signal generators. For very low frequencies, the 60-Hz power-line frequency is used as a standard in most localities. This frequency is accurately controlled by the power companies and can be used as a frequency standard. The medium AF range can be covered by comparison with the 440- and 600-Hz audio modulation on WWV transmissions. It is possible to calibrate over a 10:1 range both upward and downward from these frequencies and thus cover the audio ranges useful for voice communications.

An oscilloscope has both a horizontal and a vertical amplifier (in addition to the horizontal sweep-oscillator circuitry). This is desirable because it is convenient to have a means for adjusting the voltages applied to the deflection plates to secure a suitable pattern size. Several hundred volts is usually required for full-scale deflection in either the vertical or horizontal direction, but the current required is usually somewhere in the microampere range. Thus, the actual power needed for full-scale deflection is extremely low.

One important limitation of an oscilloscope is the frequency response of the scope deflection amplifiers. Scopes with a horizontal sweep rate that is limited to audio frequencies are relatively inexpensive and serve many amateur needs. Increasing the bandwidth of the horizontal and vertical amplifier, and increasing the sweep-oscillator frequency, will increase the useful frequency range of the scope. Of course, it is a distinct advantage for amateurs to have a scope that will handle RF through his or her most-used range; that is, to 30 MHz, to 150 MHz or even higher. Scopes with such RF capabilities are usually expensive, however.

One way to extend the useful frequency range of a narrow-bandwidth oscilloscope is with an adapter circuit. The idea is to use a mixer and an oscilloscope set to a frequency that will provide a sum or difference frequency within the useful range of your scope. With a 25-MHz oscillator, for example, you can display signals in the 20- to 30-MHz range on a 5-MHz scope. Other oscillator frequencies will enable you to display a signal with almost any desired frequency. This will not improve the transient-signal response of the narrowband scope, however.

Other limitations amateurs should be aware of are the stability of the scope sweep-frequency oscillator, absence of a horizontal deflection-plate amplifier (which would make it impossible to produce a Lissajous pattern), the ease of varying the horizontal sweep rate, the upper and lower limits of the available sweep rates, and the amount of voltage required for full-scale deflection of the electron beam. All of these factors will affect the frequency response and stability of a scope.

[Now turn to the FCC examination questions in Chapter 10 with numbers that begin 4AD-3, and study them before proceeding. Review this section as needed.]

ELECTROMAGNETIC COMPATIBILITY

We live in a world surrounded by *electromagnetic radiation*, including frequencies that range from just above dc through visible light! There is an even wider variety of electronic devices that can respond to this radiation in some way. Some devices are designed to receive specific frequencies, and others are not supposed to respond to electromagnetic waves at all. Some electronic devices generate radio signals either as their main function or as an incidental part of their operation. Very often, some of this equipment will not function as it was intended to, and radio interference will result.

As an Amateur Radio operator, you may cause interference to others, you may be interferred with or you may be blamed for interference that you do not think you could be causing. This section deals with a few of those cases, and should help you understand some of the interference, as well as prepare you to pass your Advanced class license exam.

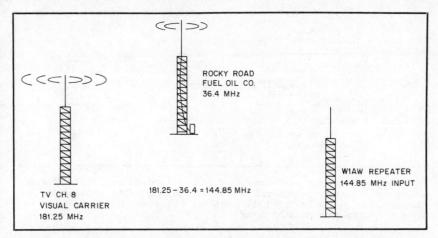

Fig. 4-8 — This diagram shows a potential intermod situation. It is possible for some of the channel 8 carrier signal at 181.25 MHz to mix with the 36.4-MHz Rocky Road carrier in the output stage of the Rocky Road transmitter. This mixing produces intermodulation distortion products at the sum and difference frequencies (181 + 36.4 = 216.4 MHz, 181 − 36.4 = 144.85 MHz). One of these intermod products is the input frequency of the W1AW repeater, and may key up the repeater. An isolator or circulator (see text) should be installed at the Rocky Road transmitter to solve this problem.

Intermodulation Distortion

Intermodulation distortion (IMD) or intermodulation interference occurs when signals from several transmitters, each operating on a different frequency, are mixed in a nonlinear receiver or transmitter stage. This produces signals at mixing products and causes severe interference in a nearby receiver. Harmonics can also be generated in the nonlinear stage, and those frequencies will add to the possible mixing combinations. The intermod, as it is called, is transmitted along with the desired signal.

For example, suppose an amateur repeater receives on 144.85 MHz. Nearby are relatively powerful nonamateur transmitters operating on 181.25 MHz and on 36.4 MHz (Fig. 4-8). Neither of these frequencies is harmonically related to 144.85. The difference between the frequencies of the two nonamateur transmitters, however, is 144.85 MHz. If both these nonamateur transmitters are transmitting at the same time, the difference frequency could be picked up by the amateur repeater and retransmitted over its own antenna, causing severe intermod.

Every transmitter operator is morally and legally required to ensure that their station emits only clean signals. The problem often is that the operators of the other transmitter do not know their equipment is causing intermod with your club's repeater. When they become aware of it, they can take several steps to eliminate the problem.

Push-pull amplifiers are quite effective in eliminating even-numbered harmonics, and it is often the second harmonic of the frequency that is causing intermod. Another solution would be to use a linear (class-AB1) power amplifier instead of a class-C amplifier. (Linear amplifiers present fewer intermod problems than do nonlinear amplifiers.) Should this remedy be impractical, at the very least, the class-C amplifier should be operated with the minimum grid current possible for efficient operation. (Chapter 7 contains detailed explanations of all these types of amplifiers and how they work.) All stages in the offending transmitter should be neutralized, to eliminate parasitic oscillations. (Low-pass and band-pass filters usually are ineffective in reducing intermod problems, because at VHF and UHF frequencies they are seldom sharp enough to suppress the offending signal without also weakening the wanted one.)

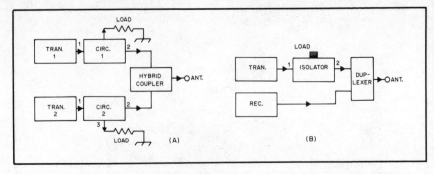

Fig. 4-9 — A block diagram showing the use of circulators and an isolator. Two circulators may be used to connect one antenna to two transmitters, as in A. Often, duplexers also are used, along with the circulators. In B, an isolator is placed between the transmitter and the duplexer to reduce intermodulation products.

Two other devices that usually are highly effective in eliminating intermod are *isolators* and *circulators*. A terminated circulator is a precisely engineered ferrite component that functions like a one-way valve. It is a three-port device that can combine two or more transmitters for operation on one antenna. Very little transmitter energy is lost as RF travels to the antenna, but a considerable loss is imposed on any energy coming down the feed line to the transmitter. Thus, the circulator effectively reduces intermod problems. Another advantage to its use is that it provides a matched load at the transmitter output regardless of what the antenna-system SWR might be.

A two-port device called an isolator also helps reduce intermod; this device incorporates a built-in termination or load. Fig. 4-9 illustrates how circulators or isolators may be included in your repeater system.

Circulators and isolators are available for 144- and 450-MHz use. They come in power levels up to a few hundred watts. Typical bandwidth for a 150-MHz unit is 3 MHz. Insertion loss is roughly 0.5 dB, and rejection of unwanted energy coming down the feeder is usually 20 to 28 dB. The 450-MHz types have greater bandwidth — about 20 MHz — but otherwise perform about the same as the 150-MHz versions. Isolators and circulators are rather expensive, but, when needed, they are worth considerably more than their price.

Intermod, of course, is not limited to repeaters. Anywhere two relatively powerful and close-by transmitter fundamental-frequency outputs or their harmonics combine to create a sum or a difference signal at the frequency on which any other transmitter or receiver is operating, an intermod problem can develop.

A two-tone IMD test on a receiver is a measure of the range of signals that can be tolerated at the receiver input before it begins to generate spurious mixing products. This specification gives an idea of how the receiver will perform in the presence of strong signals.

Another IMD topic has to do with transmitter spectral-output purity. When several audio signals are mixed with the carrier signal to generate the modulated signal, spurious signals will also be produced. These are normally reduced by filtering after the mixer, but their strength will depend on the level of the signals being mixed, among other things, and they will be present in the transmitter output to some extent. You can perform a transmitter two-tone test by putting two equal-amplitude audio tones into the microphone circuit. Then you can view the transmitter output on an oscilloscope (or spectrum analyzer) to get an indication of how linear the amplifier stages are. This is important if you are to be sure that your transmitted signal is clean.

[Before proceeding, turn to the FCC examination questions in Chapter 10 with numbers that begin 4AD-4. Review this section as needed.]

Receiver Desensitization

Another problem often encountered is *desensitization* of the receiver, almost invariably by a strong signal from a nearby transmitter (regardless of that transmitter's frequency). This signal so overloads the receiver that it becomes relatively insensitive to the signals it is supposed to receive. In the case of a desensitized repeater receiver, the offending transmitter is often that of the repeater itself. The key to eliminating desensitization is isolation — that is, isolating the receiver from any transmitters that might be causing or contributing to the problem.

Obviously, the repeater receiver and transmitter must be carefully shielded from each other. Such shielding usually involves physically separating the receiver from the transmitter and enclosing each in a metal box. All connections between the two boxes must be carefully shielded. Double-shielded coaxial cable is recommended. Separation of the transmitting and receiving antennas also is helpful, but often is impractical.

When both receiver and transmitter must use the same antenna, a series of *cavities* called a *duplexer* is employed to provide the isolation. The duplexer acts as a notch filter at the receive frequency to effectively attenuate the transmitter signal at the receiver. Total attenuation can easily be as much as 35 dB.

Often, additional capacitance is placed across the transmitter tank circuits to raise the circuit Q and make the bandwidth as small as possible. This reduces the amount of *white noise* generated by the transmitter that reaches the receiver. White noise is a constant hissing, buzzing sound in a receiver that is characteristic of unsquelched FM.

If the desensitization is the result of powerful nearby transmitters not operating on the same band, however, the use of low-pass, band-pass or high-pass filters might solve the problem. Which kind of filter is used depends on the frequency of the offending transmitter. If it is a commercial AM transmitter, a high-pass filter on the repeater receiver input might solve the problem. If not, rearrangement of the receiver antenna might also be necessary. If the problem should be a nearby commercial UHF television transmitter, a low-pass filter on the receiver input probably would help, assuming the receiver input frequency is lower than the TV frequency. If the problem is caused by a nearby amateur transmitter on the same band as the receiver, the best solution is to use a high-dynamic-range front end in your receiver.

[Go to the FCC examination questions with numbers that begin 4AD-5, and study them before proceeding. Review this section as needed.]

Cross Modulation

Cross modulation often results in one or more of several kinds of RFI. For instance, if a TV receiver is overloaded by a strong nearby signal, the result can be cross modulation. If a 14-MHz amateur signal mixes with the signal from a 92-MHz commercial FM station to produce a beat at 78 MHz, interference to VHF television channel 5 can result. The 14-MHz signal also might mix with that of a TV station on channel 5 to cause TVI on channel 3. Neither of these channels is harmonically related to 14 MHz, and in each case, both signals must be on the air for the interference to occur. Eliminating the unwanted signal at the TV receiver will cure the interference problem.

Many combinations of this type can occur, depending on the band in use and on the local frequency assignments to FM and TV stations. The interfering frequency is equal to the amateur fundamental frequency either added to or subtracted from the frequency of some local station. Should the amateur station unintentionally (but nevertheless illegally) be emitting a strong harmonic of its fundamental frequency, this harmonic also can combine with the signal from another transmitter to produce cross modulation. When interference occurs on a TV channel or a commercial frequency that is not harmonically related to the amateur transmitting frequency, the possibility of cross modulation should be investigated.

Yet another type of cross-modulation can occur with amateur phone transmitters. Occasionally a voice from an amateur transmission is heard whenever a broadcast receiver is tuned to a station, but there is no interference when tuning between stations. This type of cross modulation results from rectification of the amateur signal in one of the early receiver stages. Receivers that are susceptible to this kind of trouble usually also get a similar type of interference if there is a strong local BC station and the receiver is tuned to some other station.

The remedy for cross-modulation in the receiver is the same as for images and oscillator-harmonic response — reduce the strength of the amateur signal at the receiver by using a filter. You may also be able to reduce the interference by tuning your receiver slightly off frequency. The interference level should drop, while the desired signal is still strong.

The trouble is not always in the receiver. Cross modulation can occur in any nearby rectifying circuit such as a poor contact in water or steam piping, rain gutters, and other conductors in the strong field of the transmitting antenna, external to both receiver and transmitter. Locating the cause may be difficult. It is best attempted with a battery-operated portable BC receiver. The receiver should be used as a probe to find the spot where the interference is most intense. When such a spot is located, inspection of the metal structures in the vicinity may indicate the cause. The remedy is to make a good electrical bond between the two conductors, eliminating the rectification.

[Turn to Chapter 10 and study the FCC examination questions with numbers that begin 4AD-6. Review this section as needed.]

Capture Effect

One of the most notable differences between an amplitude-modulated (AM) receiver and a frequency-modulated (FM) receiver is how noise and interference affect an incoming signal.

From the time of the first spark transmitter, "rotten QRM" has been a major problem for amateurs. The limiter and discriminator stages in an FM receiver can eliminate most of the impulse-type noise, except any noise that has frequency-modulation characteristics. For good noise suppression, the receiver IF system and detector phase tuning must be accurately aligned.

FM receivers perform quite differently from AM, SSB and CW receivers when QRM is present, exhibiting a characteristic known as the *capture effect*. The loudest signal received, even if it is only two or three times stronger than other signals on the same frequency, will be the only signal demodulated. On the other hand, an S9 AM, SSB or CW signal suffers noticeable interference from an S2 signal. Capture effect can be an advantage if you are trying to receive a strong station and there are weaker stations on the same frequency. At the same time, this phenomenon will prevent you from receiving one of the weaker signals if that is your desire.

[Before proceeding to Chapter 5, turn to Chapter 10 and study the FCC examination questions with numbers that begin 4AD-7. Review this section as needed.]

Key Words

Apparent power — The product of the RMS current and voltage values in a circuit, without consideration of the phase angle between them.

Average power — The product of the RMS current and voltage values associated with a purely resistive circuit, equal to one half the peak power when the applied voltage is a sine wave.

Back EMF — An opposing electromotive force (voltage) produced by a changing current in a coil. It can be equal to the applied EMF under some conditions.

Bandwidth — A frequency range over which the response of a tuned circuit will be within certain limits.

Decibel — One tenth of a bel, denoting a logarithm of the ratio of two power levels.
 dB = 10 log (P2/P1)

Effective radiated power (ERP) — The relative amount of power radiated in a specific direction from an antenna, taking system gains and losses into account.

Electric field — A region through which an electric force will act on an electrically charged object.

Field — Any of the invisible forces in nature, such as gravity, electric force or magnetic force. This term also refers to the region of space through which these forces act.

Half-power points — Those points on the response curve of a resonant circuit where the power is one half its value at resonance.

Joule — The unit of energy in the metric system of measure.

Magnetic field — A region through which a magnetic force will act on a magnetic object.

Parallel-resonant circuit — A circuit including a capacitor, an inductor and sometimes a resistor, connected in parallel, and in which the inductive and capacitive reactances are equal at the applied-signal frequency. The circuit impedance is a maximum, and the current is a minimum at the resonant frequency.

Peak-envelope power (PEP) — The average power of the RF envelope during a modulation peak. (Used for modulated RF signals.)

Peak power — The product of peak voltage and peak current in a resistive circuit. (Used with sine-wave signals.)

Phase — A representation of the relative time or space between two points on a waveform, or between related points on different waveforms.

Phase angle — If one complete cycle of a waveform is divided into 360 equal parts, then the phase relationship between two points or two waves can be expressed as an angle.

Power — The time rate of transferring or transforming energy, or the rate at which work is done.

Power factor — The ratio of real power to apparent power in a circuit. Also calculated as the cosine of the phase angle between current and voltage in a circuit.

Q — A quality factor describing how closely a practical coil or capacitor approaches the characteristics of an ideal component.

Reactive power — The apparent power in an inductor or capacitor. The product of RMS current through a reactive component and the RMS voltage across it. Also called wattless power.

Real power — The actual power dissipated in a circuit, calculated to be the product of the apparent power times the phase angle between the voltage and current.

Rectangular coordinates — A graphical system used to represent length and direction of physical quantities for the purpose of finding other unknown quantities.

Resonant frequency — That frequency at which a circuit including capacitors and inductors presents a purely resistive impedance. The inductive reactance in the circuit is equal to the capacitive reactance.

Series-resonant circuit — A circuit including a capacitor, an inductor and sometimes a resistor, connected in series, and in which the inductive and capacitive reactances are equal at the applied-signal frequency. The circuit impedance is at a minimum, and the current is a maximum at the resonant frequency.

Skin effect — A condition in which ac flows in the outer portions of a conductor. The higher the signal frequency, the less the electric and magnetic fields penetrate the conductor, and the smaller the effective area of a given wire for carrying the electrons.

Thevenin's Theorem — Any combination of voltage sources and impedances, no matter how complex, can be replaced by a single voltage source and a single impedance that will present the same voltage and current to a load circuit.

Chapter 5

Electrical Principles

To pass the Technician or General class exam, you studied basic electronic principles dealing with dc circuits, along with the ac theory for some elementary components such as resistors, capacitors and inductors. The Advanced class examination will emphasize ac circuit theory. You will be expected to have a good working knowledge of reactance and impedance, be able to perform power calculations, and to know what reactive power and power factor are. The Element 4A exam also covers energy storage in capacitors and inductors, resonant circuits, phase angle and circuit Q. Thevenin's Theorem is a powerful tool that you can use to simplify the solution to many circuit problems. You will likely have one or more questions on your exam about using this important electronics rule.

This chapter explains all of these topics so that you will understand how to perform the required calculations and will know the information covered by the exam questions. There are many sample problems worked out in the text. Follow those explanations carefully, being sure to work out the arithmetic and manipulate the equations as indicated. As you complete each section of text, you will be instructed to turn to Chapter 10 and study the appropriate group of examination questions. If you have difficulty with any of the questions, be sure to review the material in this chapter. If you would like to delve deeper into the theory behind the topics covered here, you may want to read some more-advanced texts. *The ARRL Handbook* is always on the recommended reading list, of course. There are many good books about electronics theory and principles, and by reading some of the material in another book you may get a different viewpoint or an explanation that is easier for you to understand.

ELECTRICAL ENERGY

Before you can understand what electrical energy is, you must know some important definitions. Let's use some simple examples from your everyday experience to build those definitions. Pick up a stone, and carry it to an upstairs window of your house. You are doing work against the earth's gravity as you move the stone farther away from the earth's surface. Gravity is an invisible force that the earth exerts on the stone. There are many invisible forces in nature, and these are often referred to as fields. (No, this is not the same as a farmer's field!) What we mean here is a region of space through which a force acts without actual contact. When you pick up the stone, there is physical contact, so you are not exerting an invisible force — you are not a field, then. The earth's gravitational field pulls every object toward the center of the earth.

Okay, so how much work did you do on the stone? Well, you have to multiply the distance you moved it through the gravitational field (let's say to a height of 10 feet above the earth) times the force you had to exert, which was equal to the force of gravity on the stone (say 1 pound). So you have done 10 foot-pounds of work against the gravitational field. Now place the stone on a windowsill. By doing work on the stone, you have stored some energy in it. That energy is equal to the amount of work that you did, and is called potential energy. In effect, you are storing the energy by the stone's position in the earth's gravitational field. If you push the stone out the window, it will fall back to the earth, and while it is falling, the stored potential energy is being converted to kinetic energy, or energy of motion.

In electronics we are interested mainly in two types of invisible forces, or fields. Those are the *electric field* and the *magnetic field* that make up an electromagnetic wave. (A wave of this type is a field in motion.) We can store electrical potential energy as a voltage in an electric field and as a current in a magnetic field. In either case, that potential energy can be released in the form of an electric current in the circuit.

Storing Energy in an Electric Field

You can store energy in a capacitor by applying a dc voltage across the terminals. There will be an instantaneous inrush of current to charge the capacitor plates. The only thing limiting the current at the instant that the voltage is connected is any resistance that there may be in the circuit. (Of course there will always be some resistance in the wires connecting the components, and in the components themselves.) The capacitor builds up an electric charge as one set of plates accumulates an excess of electrons and the other set loses an equal number. The voltage across the capacitor rises as this charge builds up. Eventually, the voltage at the capacitor terminals is equal to the source voltage, and the current stops. If the voltage source is disconnected, the capacitor will remain charged to that voltage. The charge will stay on the capacitor plates as long as there is no path for the electrons to travel from one plate to the other.

At this point we should be careful to point out that our discussion in this section deals with ideal components. We are thinking of resistors that have no stray capacitance or inductance associated with the leads or composition of the resistor itself. Ideal capacitors exhibit no losses, and there is no resistance in the leads or capacitor plates. Ideal inductors are made of wire that has no resistance, and there is no stray capacitance between turns. Of course, in practice we do not have ideal components, so the conditions described here may be modified a bit in real-life circuits. Even so, components can come pretty close to the ideal conditions. For example, a capacitor with very low leakage will hold a charge for days or even weeks.

Stored electric potential energy produces an electric field in the capacitor. Since the charge is not moving, this field is sometimes called an electrostatic field. If a resistor or some other circuit is connected across the capacitor terminals, that field will return the stored energy by creating a current in the circuit.

Fig. 5-1 illustrates a simple circuit for charging and discharging a capacitor, depending on the switch setting. The electric energy will be converted to heat energy in a resistor, or into other forms of energy, such as sound energy in a speaker. If the capacitance is high, and

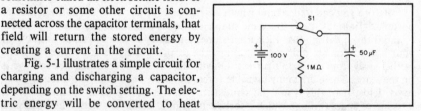

Fig. 5-1 — A simple circuit for charging a capacitor and then discharging it through a resistor.

you connect a large-value resistance across the terminals, it may take a long time for all the energy to be dissipated and the capacitor voltage to drop to zero. On the other hand, by touching a wire across the terminals to short circuit the capacitor, you can discharge it very quickly. If you short a capacitor of several microfarads that has been charged to several hundred volts, you can produce quite a spark. If your skin touches the terminals instead of a piece of wire, you can get a dangerous shock. Large-value filter capacitors in a power supply have bleeder resistors connected across them to drain this charge when the supply is turned off.

The basic unit for expressing energy in the metric system, which is the system used to express all common electrical units, is the *joule* (pronounced with a long u sound, similar to jewel). Electrical energy (or work) is the product of the force needed to move a single electron through an electric potential, times the distance it moves. So if it takes a force of 1 newton to move an electron 1 meter through an electric field, 1 joule of work is done, and 1 joule of energy is stored in the electron.

We are not usually concerned with single electrons, however. We want large numbers of electrons to flow in a circuit. You probably remember from your Novice days that if one coulomb of electrons (6.24×10^{18} or 6 quintillion, 240 quadrillion electrons) flow past a point in one second, then we have a current of one ampere. So even a current of one microampere represents over 6 trillion electrons flowing past a point per second! If 1 joule of work is done on an entire coulomb of electrons, then we say the electric potential is 1 volt. Stating this in mathematical terms, we can write:

$$1 \text{ volt} = \frac{1 \text{ joule}}{1 \text{ coulomb}} \qquad \text{(Eq. 5-1)}$$

Factors that Determine Capacitance

At this point it may also be helpful to recall a few simple facts about capacitors from the Technician/General class theory. A capacitor is simply two conductors separated by some insulating material. The capacitance increases with plate size and decreases with increasing distance between them. Capacitance also depends on the type of material used as the insulator. The insulating properties of the insulator are related to the dielectric constant of the particular substance. The dielectric constant compares all materials to the insulating properties of a vacuum, which is 1. The insulating properties of air are nearly the same as free space, so air has a dielectric constant of 1. Table 5-1 lists some common insulating materials and their dielectric constants.

Table 5-1
Dielectric Constants for Some Common Insulators

Material	Dielectric Constant
Air	1.0
Bakelite	4.4-5.4
Glass, window	7.6-8
Glass, Pyrex®	4.8
Mica	5.4
Plexiglas®	2.8
Polyethylene	2.3
Polystyrene	2.6
Teflon®	2.1

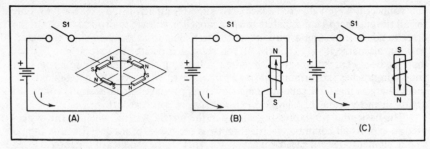

Fig. 5-2 — Simple circuits illustrating the magnetic field around a straight wire (A) and two coils wound in opposite directions (B and C).

The actual work done in charging a capacitor, or the energy stored in the charged capacitor, can be found by:

$$W = \frac{V^2C}{2}$$ (Eq. 5-2)

where
   W = work or energy in joules (or watt-seconds)
   V = potential in volts
   C = capacitance in farads

Storing Energy in a Magnetic Field

When electrons flow through a conductor, a magnetic field is produced. This can be demonstrated by bringing a compass near a current-carrying wire and watching the needle deflect. Fig. 5-2A shows the magnetic field around a wire connected to a battery. If the wire is wound into a coil, so that the fields from adjacent turns will add together, then a much stronger magnetic field can be produced. The direction of the field, which points to the magnetic north pole, can be found using a "left hand rule." For a straight wire, point the thumb of your left hand in the direction of the electron flow and your fingers curl in the direction of the north pole of the magnetic field around the wire. In a similar manner, if you curl the fingers of your left hand around the coil in the direction of electron flow, your thumb points in the direction of the north pole of the field. Parts B and C of Fig. 5-2 illustrate the fields around two coils wound in opposite directions.

The strength of the magnetic field depends on the amount of current, and is stronger when more current flows. Electrical energy from the voltage source is transferred to the magnetic field in the process of creating the field. So we are storing energy by building up a magnetic field, and that means work must be done against some opposing force. That opposing force is the result of a voltage induced in the circuit whenever the magnetic field (or current) is changing. If the current remains constant, then the magnetic field remains a constant, and there is no more energy being stored.

When you first connect a dc source to a coil of wire, a current begins to flow, and a magnetic field begins to build up. The field is changing very rapidly at that time, so a large opposing voltage is created, preventing a large current from flowing. As a maximum amount of energy is stored and the magnetic field reaches its strongest value, the opposing voltage will decrease to zero, so the current increases gradually to a maximum value. That maximum value is limited only by the resistance of the wire in the coil, and by any internal resistance of the voltage source. If the current decreases, then a voltage is induced in the wire that will try to prevent the decrease. The stored energy is being returned to the circuit in this case.

The work done in producing a magnetic field in an inductor (or the energy stored in the field) can be found by:

$$W = \frac{I^2 L}{2}$$

(Eq. 5-3)

where

W = work in joules (or watt-seconds)
I = current through the coil in amperes
L = inductance in henrys

This induced voltage or EMF is sometimes called a back EMF, since it is always in a direction to oppose any change in the amount of current. When the switch in the circuit of Fig. 5-2B is first closed, this back EMF will prevent a sudden surge of current through the coil. Notice that this is just the opposite of the condition when a capacitor is charging. Likewise, if you open the switch to break the circuit, a back EMF will be produced in the opposite direction. This time the EMF tries to keep the current going, again preventing any sudden change in the magnetic-field strength.

The magnitude of the induced back EMF depends on how rapidly the current is changing. If you have a strong magnetic field built up in a coil, and then suddenly break the circuit at some point, a large voltage is induced in the coil, which tries to maintain the current. It is quite common to have a spark jump across the switch contacts as they open. In fact, this is exactly the principle used in the induction coil of an automobile. A large current flows through the coil, then at the proper instant a set of contacts open, inducing a much larger voltage in a second coil. The voltage in this second coil causes a spark to jump across a spark-plug gap, and the gasoline in the cylinder explodes.

[Turn to Chapter 10 at this time, and study FCC examination questions with numbers that begin 4AE-4. Review this section as needed.]

PHASE ANGLE BETWEEN CURRENT AND VOLTAGE

Now that you understand how capacitors and inductors store energy in electric and magnetic fields, we can learn how those devices react when an alternating voltage is applied to their terminals. While the resistance of a pure resistor does not vary with the frequency, you already know that the reactance of both a coil and a capacitor do change with frequency. That should tell us that coils and capacitors will behave differently with an ac voltage than a dc one. Remember that a pure inductor does not impede direct current, but does impede alternating current, and the higher the frequency the more it opposes the current. Also a pure capacitor will not allow dc to pass through it, but will hamper ac less and less as the frequency increases.

To understand the variations in voltage and current through inductors and capacitors, we must look at the amplitudes of these ac signals at certain instants of time. The relationship between the current and voltage waveforms at a specific instant is called the *phase*. Phase essentially means time, or a time interval between when one event occurs and the instant when a second, related, event takes place. The event that occurs first is said to lead the second, while the second event lags the first.

Since each ac cycle takes exactly the same time as any other cycle of the same frequency, we can use the cycle as a basic time unit. This makes the phase measurement independent of the waveform frequency. If two or more different frequencies are being considered, phase measurements are usually made with respect to the lowest frequency.

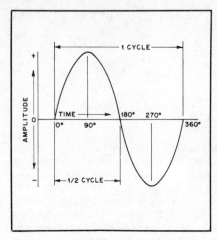

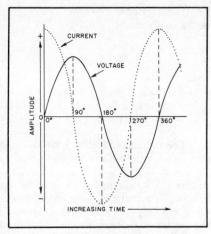

Fig. 5-3 — An ac cycle is divided into 360° that are used as a measure of time or phase.

Fig. 5-4 — Voltage and current phase relationships when an alternating voltage is applied to a capacitor.

It is convenient to relate one complete cycle of the wave to a circle, and to divide the cycle into 360 equal parts or degrees. So a phase measurement is usually specified as an angle. In fact, we often refer to the *phase angle* between two waveforms. Fig. 5-3 shows one complete cycle of a sine-wave voltage or current, with the wave broken into four quarters of 90° each.

AC Through a Capacitor

As soon as a voltage is applied across the plates of an ideal capacitor there is a sudden inrush of current as the capacitor begins to charge. That current tapers off as the capacitor is charged to the full value of applied voltage. By the time the applied voltage is reaching a maximum, the capacitor is also reaching full charge, and so the current through the capacitor goes to zero. A maximum amount of energy has been stored in the electric field of the capacitor at this point.

Fig. 5-4 graphs the relative current and voltage amplitudes over time. The two lines are shown differently to help you distinguish between them; the scale is not intended to show specific current or voltage values. When the applied voltage passes the peak and begins to decrease, the capacitor starts returning some of its stored energy to the circuit. Electrons are now flowing in a direction opposite to the direction they were flowing when the capacitor was charging. By the time the applied voltage reaches zero, the capacitor has returned all of its stored energy to the circuit, and the reverse-direction current is a maximum value. Now the applied voltage direction has reversed, and a large charging current is applied to the capacitor, which decreases to zero by the time the applied voltage reaches a maximum value. The second half of the applied-voltage cycle is exactly the same as the first half, except the relative directions of current and voltage are reversed.

Study Fig. 5-4 to understand the current and voltage relationships for a capacitor over an entire cycle. Notice that the current reaches each point on a cycle 90° ahead of the applied-voltage waveform. We say that the current through a capacitor leads the applied voltage by 90°. You could also say that the voltage applied to a capacitor lags the current through it by 90°. Notice also that the applied-voltage waveform and the stored electric-field waveform are in phase — that is, similar points on those waveforms occur at the same instant (Fig. 5-5).

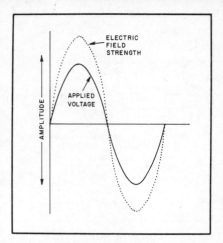

Fig. 5-5 — The applied voltage and electric field in a capacitor are in phase.

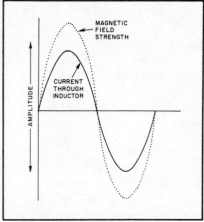

Fig. 5-6 — The current through an inductor is in phase with the magnetic field strength.

AC Through an Inductor

The situation with ac through an inductor is a little more difficult to understand than the capacitor case. As we go through the conditions for a single current cycle, study the graph in Fig. 5-6. That should make it easier to follow the changing conditions as they are described.

Let's apply an alternating current to an ideal inductor and observe what happens. At the instant when the current is zero and starts increasing in a positive direction, a magnetic field will start to build up around the coil, storing the applied energy. The current is increasing at a maximum rate, so the magnetic field strength is increasing at a maximum rate. As the current through the coil reaches a positive peak and begins to decrease, a maximum amount of energy has been stored in the magnetic field, and then the coil will begin to return energy to the circuit as the magnetic field collapses. When the current is crossing zero on the way down, it is again changing at a maximum rate, so the magnetic field is also changing at a maximum rate. As the current direction changes, it will begin to build a magnetic field in the direction opposite to the first field, and energy is being stored once again.

When the current reaches a maximum negative value, a maximum magnetic field has been built up, and then the stored energy begins to return to the circuit again as the current increases toward zero. The second half of the cycle is the same as the first, but with all polarities reversed. You should realize by now that the current and the magnetic field in an inductor are in phase — similar points on both waveforms occur at the same instant.

In the section on storing energy in an inductor, you learned about the back EMF that is induced in the coil. That EMF is greatest when the magnetic field is changing the fastest. Furthermore, it is in a direction that opposes the change in current or magnetic-field strength. So when the current is crossing zero on the way to a positive peak, the induced EMF is at its greatest negative value. When the current is at the positive peak, the back EMF is zero, and so on. Fig. 5-7 shows the phase relationship between the current through the inductor and the back EMF across it.

Since we know that the back EMF opposes the effects of an applied voltage, we can draw in the waveform for applied voltage as shown in Fig. 5-7. The applied

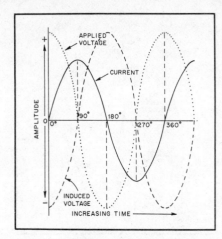

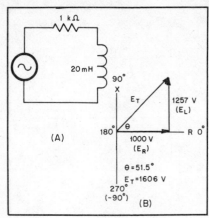

Fig. 5-7 — Phase relationships between voltage and current when an alternating voltage is applied to an inductance.

Fig. 5-8 — A series RL circuit is shown at A. B shows the right triangle used to calculate the phase angle between the circuit current and voltage.

voltage is 180° out of phase with the induced voltage. From this fact, Fig. 5-7 shows that the voltage across an inductor leads the current through it, or the current lags the voltage. The phase relationship between applied voltage and current through an inductor is just the opposite from the relationship for a capacitor. A useful memory technique to remember these relationships is the little saying, "ELI the ICE man." The L and C represent the inductor and capacitor, and the E and I stand for voltage and current. Right away you can see that E comes before (leads) I in an inductor and that I comes before E in a capacitor.

Phase Angle With Real Components

Up to this point we have been talking about ideal components in discussing the phase relationships between voltage and current in inductors and capacitors. Of course, any real components will have some resistance associated with the inductance or capacitance. Since the voltage across a resistor is in phase with the current through it, the overall effect is that the phase difference between voltage and current will be less than 90° in both cases.

Solving Problems Involving Inductors

Probably the easiest way to illustrate the effect of adding resistance to the circuit is by means of a problem. Fig. 5-8A shows a simple circuit with a resistor and an inductor in series with an ac signal source. Let's pick a frequency of 10 kHz for the signal generator, and connect it to a 20-mH inductor in series with a 1-kΩ resistor. The question is, what is the phase angle between the voltage and current in this circuit? To aid our solution, draw a set of *rectangular coordinates,* and label the X (horizontal) axis R (for resistance) and the Y (vertical) axis X (for reactance). The degree indications shown on Fig. 5-8B show the standard way of relating the coordinate system to degrees around a circle. This coordinate system can represent either voltage or current, as needed. In this case, it represents voltage, since the current is the same in all parts of a series circuit. Next, we must calculate the inductive reactance of the coil:

$$X_L = 2\pi fL = 6.28 \times (10 \times 10^3 \text{ Hz}) \times (20 \times 10^{-3} \text{ H}) = 1257 \ \Omega$$

Now calculate the voltage across the inductor using Ohm's Law. Since the actual current is not important, as long as we know it is the same through each part of the circuit, choose a simple value, such as 1 A:

$$E_L = I\, X_L = 1\text{ A} \times 1257\ \Omega = 1257\text{ V}$$

The voltage across the resistor is also found easily using Ohm's Law:

$$E_R = I\, R = 1\text{ A} \times 1000\ \Omega = 1000\text{ V}$$

Wow! What a nifty trick. By assuming a current of 1 A when the actual current value is not important to the problem, we are able to eliminate a step in the solution, because the voltage can be considered equal to the reactance or resistance!

Okay, so now what? Well, we can't just add the two voltages to get the total voltage across the circuit, because they are not in phase. Remember that the voltage across an inductor leads the current, and the voltage across a resistor is in phase with the current. This means that the peak voltage across the inductor occurs 90° before the peak voltage across the resistor. On your graph, draw a line along the R (horizontal) axis to represent the 1000 V across the resistor. At the end of that line draw another line to represent the 1257 V across the inductor. Since this voltage leads the voltage in the resistor by 90°, it must be drawn vertically upward. Now complete the figure by drawing a line from the origin (the point where the two axes cross) to the end of the E_L line. This right triangle represents the solution to our problem. Fig. 5-8B shows a complete drawing of the triangle. The last line is the total voltage across the circuit, and the angle measured up from the E_R line to the E_T line is the phase angle between the voltage and current. If you use graph paper and make up a suitable scale of divisions per volt on the graph, you can actually solve the problem with no further calculation. Graphical solutions are usually not very accurate, however, and depend a great deal on how carefully you draw the lines.

Let's complete the mathematical solution to our problem. If you are not familiar with right triangles and their solutions (a branch of mathematics called trigonometry), just follow along. You should be able to pick up the techniques from a few examples, but if you continue to have trouble, go to your library and check out a book on elementary high school trigonometry. The various sides and angles are identified by their positions in relation to the right (90°) angle. The side opposite the right angle is always the longest side of a right triangle, and is called the hypotenuse. The other sides are either opposite or adjacent to the remaining angles. It is important to realize that these methods of trigonometry apply only to triangles that contain a 90° angle.

We will find the phase angle first, using the tangent function. Since the angle of interest is the one between the E_R line and the hypotenuse, E_L is the side opposite and E_R is the side adjacent to the angle. Tangent is defined as:

$$\tan \theta = \frac{\text{side opposite}}{\text{side adjacent}} \qquad \text{(Eq. 5-4)}$$

Then for our problem,

$$\tan \theta = \frac{E_L}{E_R} = \frac{1257\text{ V}}{1000\text{ V}} = 1.257$$

where θ is the angle between the E_R and E_T lines.

Now that we know the tangent of the angle, it is a simple matter to refer to a table of trigonometric functions to find what angle has a tangent of that value. If you have an electronic calculator that is capable of doing trig functions, then you simply ask it to find the angle using the inverse tangent or acrtangent function, often written $\tan^{-1}$:

$$\tan^{-1}(1.257) = 51.5°$$

If the resistance and reactance of our components had been equal, then we would have found $\tan\theta$ to be 1, and the phase angle would be 45°. If the reactance were many times larger than the resistance, then the phase angle would be close to 90°, and if the resistance were many times larger than the reactance, then the phase angle would be close to 0°.

There is one more question we would like an answer to here, and that is the total assumed voltage across the circuit, given our assumed current of 1 ampere. We could find the actual voltage in the same manner, given an actual current through the circuit. There are two common methods available to find this, so let's look at both of them. The first method uses the sine function.

$$\sin\theta = \frac{\text{side opposite}}{\text{hypotenuse}} \qquad \text{(Eq. 5-5)}$$

We know that the phase angle is 51.5°, so we find the value of the sine function for that angle (using a trig table, calculator or slide rule). Sin (51.5°) = 0.7826. By solving Eq. 5-5 for the hypotenuse, we can find the answer:

$$\text{hypotenuse} = E_T = \frac{\text{side opposite}}{\sin\theta} = \frac{E_L}{\sin(51.5°)} = \frac{1257 \text{ V}}{0.7826}$$

$$E_T = 1606 \text{ V}$$

The second method for finding E_T involves the use of an equation known as the Pythagorean Theorem (named after Pythagoras, who discovered this important relationship). The sides of a right triangle are related to each other by the equation:

$$C^2 = A^2 + B^2 \qquad \text{(Eq. 5-6)}$$

where
   C = the length of the hypotenuse
   A and B = the lengths of the other two sides.

We can solve this equation for the length of the hypotenuse, and rewrite it:

$$C = \sqrt{A^2 + B^2} \qquad \text{(Eq. 5-7)}$$

Since the side we want to find is the hypotenuse of our triangle (side E_T), we can use Eq. 5-7.

$$E_T = \sqrt{(1257 \text{ V})^2 + (1000 \text{ V})^2} = \sqrt{1580049 \text{ V}^2 + 1000000 \text{ V}^2}$$

$$E_T = \sqrt{2580049 \text{ V}^2} = 1606 \text{ V}$$

It doesn't matter which method you use to find the total voltage, so pick whichever one seems easier, then stick with it. After you have worked a few of these problems, they will begin to seem much easier. It is very important to follow an organized, systematic approach, however, or you will become easily confused, and any slight change in the wording of the problem will throw you off.

Solving Problems Involving Capacitors

Let's try another problem, to see how well you understood that solution. Fig. 5-9A shows a circuit with a capacitor instead of an inductor. For simplicity, we will keep the same signal-generator frequency and the same resistor in the circuit.

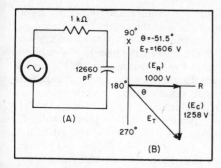

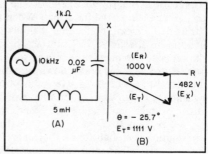

Fig. 5-9 — A series RC circuit is shown at A. B shows the right triangle used to calculate the phase angle between the circuit current and voltage.

Fig. 5-10 — A series RLC circuit is shown at A. B shows the right triangle used to calculate the phase angle between circuit current and voltage.

The 12660-pF capacitor is a strange value, but let's use it anyway, just for the sake of example.

After drawing a set of coordinates to draw the triangle on, we have to calculate the capacitive reactance.

$$X_C = \frac{1}{2 \pi fC} = \frac{1}{6.28 \times (10 \times 10^3 \text{ Hz}) \times (12660 \times 10^{-12} \text{ F})}$$

$$X_C = \frac{1}{7.95 \times 10^{-4}} = 1258 \ \Omega$$

Assume a current of 1 A and use Ohm's Law to calculate the voltages across the resistor and capacitor. $E_R = 1000$ V and $E_C = 1258$ V. Draw the resistor-voltage line on your diagram, as before. Keep in mind that the voltage across a capacitor lags the current through it, so the capacitor voltage is 90° behind the resistor voltage. Show this on your graph by drawing the capacitor-voltage line vertically downward at the end of the E_R line. Actually, we should refer to the capacitor voltage as a negative value, -1258 V, then. Complete the triangle by drawing the E_T line, and proceed with the solution as in the previous problem.

Tan $\theta = -1.258$, $\tan^{-1}(-1.258) = -51.5°$ and $E_T = 1606$ V (1607 V by the Pythagorean Theorem solution). Notice that the phase angle came out negative this time, because the capacitor voltage was taken as a negative value. This indicates that the total voltage across the circuit is 51.5° behind the current — a result of the fact that the voltage across a capacitor lags the current through it. Of course we could also say that the current leads the voltage by 51.5°, and that would mean the same thing.

Solving Problems Involving Both Inductors and Capacitors

Series Circuits

The problem illustrated in Fig. 5-10 adds a slight complication, but don't get too confused. Start out the same way we did for the first two problems, by drawing a set of coordinates, labeling the axes, and then calculating the reactance values. Since this is still a series circuit, the current will be the same through all components, so we are still interested in calculating the voltages. The inductive reactance of a 5-mH inductor with a 10-kHz signal applied to it is 314 Ω, and the capacitive reactance of a 0.02-μF capacitor with the same signal is 796 Ω. Now write a value for the assumed voltages across each component: $E_R = 1000$ V, $E_L = 314$ V and $E_C = -796$ V.

(Don't forget that the voltage across the capacitor is negative, 180° out of phase with the voltage across the inductor.) When we go to add these lines to make up our "triangle" you will notice that the lines for the inductor and capacitor voltages go in opposite directions. Just subtract them before drawing the line. In this example, the capacitor voltage is the larger, so the total reactive voltage is -482 V, and is represented by a line drawn downward on the diagram. If the inductor voltage were greater, then the total reactive voltage would be positive, and would be represented by a line drawn upward on the diagram. That just tells us if the phase angle between circuit voltage and current will be positive (upward) or negative (downward).

Okay, so now you have a triangle drawn to look like the one in Fig. 5-10B, and the solution is straightforward. Tan $\theta = -0.482$, $\tan^{-1}(-0.482) = -25.7°$ and $E_T = 1111$ V. The total voltage across our circuit lags the current by 25.7°.

Parallel Circuits

Now for the ultimate complication in this type of circuit problem! All of our work so far has been with series circuits. But what happens if the components are connected in parallel across the signal source? Let's handle this in one big jump, with a circuit including a resistor, an inductor and a capacitor all connected in parallel with our 10-kHz signal generator. Fig. 5-11 gives a circuit diagram, with component values.

Again, start by drawing a set of coordinates, labeling the axes R for resistance and X for reactance, and then calculate the inductive and capacitive reactances. $X_L = 628$ Ω and $X_C = 159$ Ω.

Now the problem becomes a little different than all the others we have done so far. Because this is a parallel circuit, we must recognize that it is the applied voltage that is the same for all components, and that the current through each will have a different phase. Of course the current through a resistor is in phase with the voltage across it, as always. The current through the capacitor leads the voltage across it, and the current through an inductor lags the voltage across it.

Since the actual value of applied voltage is not important, we can assume a value that will make our problem as easy as possible. Let's pick a value of 1000 V. Why 1000? Well, as always, Ohm's Law applies to the circuit, so we can solve it for the individual currents through each component:

$$I_R = \frac{E}{R}, \quad I_L = \frac{E}{X_L} \text{ and } I_C = \frac{E}{X_C}$$

(Eq. 5-8)

Since the largest value of resistance or reactance is 628 Ω, all of our currents will come out as whole numbers. Will the solution be the same if you pick 10 V or 1 V? Yes! In fact, it would be a good exercise for you to also work out the problem with those (or other) values of applied voltage.

With the applied voltage of 1000, $I_R = 10$ A, $I_L = -1.59$ A and $I_C = 6.29$ A. Why is the current through the inductor negative? Because it lags the voltage by 90°. The minus sign tells us that when we combine the current

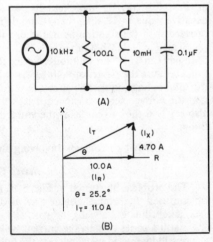

Fig. 5-11 — A parallel RLC circuit is shown at A. B shows the right triangle used to calculate the phase angle between circuit voltage and current.

through the inductor with the current through the capacitor, we have to subtract the numbers because they go in opposite directions. Whichever number is larger determines if the current is leading or lagging the voltage, and whether to draw the line up or down on your triangle. For this problem, $I_X = 4.70$ A. The current through the capacitor is larger, so the total current leads the voltage. After drawing the line for resistor current on your graph, draw a line upward to represent the capacitive current in the circuit. Complete the triangle and continue with the solution, as before. Tan $\theta = 0.470$, $\tan^{-1}(0.470) = 25.2°$ and $I_T = 11.0$ A. That's all there is to it!

How would you find the phase angle in a series or parallel circuit that contains several inductors, capacitors and resistors? Well, it really is no more difficult than any of the problems we have done so far. After calculating all the reactances, you can find an assumed voltage across the components in a series circuit or an assumed current through the components in a parallel circuit. Then the combined voltage or current is calculated by adding the values for like components (inductive, capacitive and resistive values added separately). Finally, draw a triangle with the resistive value on the horizontal axis and the reactive value on the vertical axis. Just remember that a leading voltage or current is drawn upward and a lagging voltage or current is drawn downward.

[If you worked all of these problems along with the text, you should have no trouble with the phase-angle questions on the Advanced class exam. Just to prove that to yourself, turn to Chapter 10 now. Study FCC examination questions with numbers that begin 4AE-6. Review the examples in this section if needed.]

RESONANT CIRCUITS

With all of the problems so far, we have used inductor and capacitor values that give different inductive and capacitive reactances. So there was always a bigger voltage across one of the series components or a larger current through one of the parallel ones. But you may have wondered about the condition if both reactances turned out to be equal, with equal (but opposite) voltages or currents. You have probably realized that in a series circuit with an inductor and a capacitor, if the inductive reactance is equal to the capacitive reactance, then the voltage drop across each component is the same, but they are 180° out of phase. The two values cancel, and the only remaining voltage drop is across any resistance in the circuit.

It is a common practice to say that the two reactances cancel in this case, and to talk about inductive reactance as a positive value and capacitive reactance as a negative value. Just keep in mind that this terminology is a simplification that applies only to series circuits, and comes about because the voltage across the inductor is positive or leads the current and that the voltage across the capacitor is negative or lags the current.

With a parallel circuit, if the inductive and capacitive reactances are equal, then the currents through the components will be equal, but 180° out of phase. The two currents cancel, and the only current in the circuit is a result of a parallel resistance. Here again, it is common practice to say that the reactances cancel. In the parallel circuit, however, it is the inductive component that is considered to be negative and the capacitive component is considered to be positive. Of course that terminology comes from the fact that the current through a parallel inductor lags the applied voltage and the current through a parallel capacitor leads the applied voltage.

Whether the components are connected in series or parallel, we say the circuit is resonant at the one frequency at which the inductive-reactance value is the same as the capacitive-reactance value. Remember that inductive reactance increases as the frequency increases and that capacitive reactance decreases as frequency increases. Fig. 5-12 is a graph of inductive and capacitive reactances for two general components.

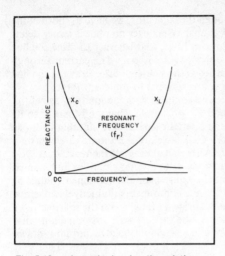

Fig. 5-12 — A graph showing the relative change in inductive reactance and capacitive reactance as the frequency increases. For a specific inductor and a specific capacitor, the point where the two curves cross ($X_L = X_C$) is called the resonant frequency.

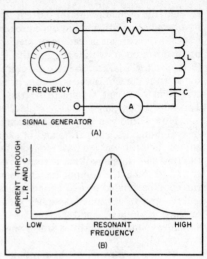

Fig. 5-13 — A series-connected LC or RLC circuit behaves like a very low resistance at the resonant frequency. Therefore, at resonance, the current passing through the components reaches a peak.

The exact frequency scale is unimportant, as is the exact reactance scale. The two lines cross at only one point, and that point represents the *resonant frequency* of a circuit using those two components. Every combination of a capacitor and an inductor will be resonant at some frequency.

Since resonance occurs when the reactances are equal, we can derive an equation to calculate the resonant frequency of any capacitor-inductor pair:

$$X_L = 2\pi f_r L = X_C = \frac{1}{2\pi f_r C}$$

$$1 = \frac{1}{(2\pi f_r L)(2\pi f_r C)}$$

$$f_r^2 = \frac{1}{(2\pi)^2\, LC}$$

$$f_r = \frac{1}{2\pi\, \sqrt{LC}} \qquad\qquad \text{(Eq. 5-9)}$$

Series-Resonant Circuits

If we connect a signal source with an output at the resonant frequency of a series RLC circuit, as shown in Fig. 5-13, the current through the circuit will cause a voltage drop to appear across each component. Because the voltage drops across the inductor and capacitor are 180° out of phase, they can be many times as large as the voltage applied to the circuit. In fact, those voltages are sometimes at least 10 times as large, and may be as much as a few hundred times as large as the applied voltage in a practical circuit. With perfect components and no resistance in the circuit, there would be nothing to restrict the current in the circuit. An ideal *series-resonant circuit*, then, "looks like" a short circuit to the signal generator. There is always some resistance in a circuit, but if the total resistance is small, the current will be large, by Ohm's Law. See Fig. 5-13B. The large current that results from this condition produces large, but equal, voltage drops across each reactance. The phase relationship between the voltages across each reactance means that the voltage across the coil reaches a positive peak at the same time that the voltage across the capacitor reaches a negative peak. We can say that the impedance of an ideal series-resonant circuit is zero, and is approximately equal to the circuit resistance in an RLC circuit. The main purpose of the series resistor is to prevent the circuit from overloading the generator.

We can think of the resonance condition in another way, which may help you understand these conditions a little better. The large voltages across the reactances develop because of energy stored in the electric and magnetic fields associated with the components. The energy going into the magnetic field on one half cycle is coming out of the electric field of the capacitor. Then, when all of the energy has been transferred to the magnetic field, it is returned to the circuit and is stored in the electric field of the capacitor again. A large amount of energy can be handed back and forth between the inductor and the capacitor without the source supplying any additional amount. The source only has to supply the actual power dissipated in the resistance of imperfect inductors and capacitors plus what is used by the resistor.

Let's see how you do with a series-resonant-circuit problem. What frequency should the signal generator in Fig. 5-13 be tuned to for resonance if the resistor is 10 kΩ, the coil is a 6.28-μH inductor and the capacitor has a value of 20 pF? Probably the biggest stumbling block on these problems will be remembering to convert the inductor value to henrys and the capacitor value to farads. After you have done that, use Eq. 5-9 to calculate the resonant frequency.

$6.28 \ \mu H = 6.28 \times 10^{-6} \ H = 0.00000628 \ H$

$20 \ pF = 20 \times 10^{-12} \ F = 0.000000000020 \ F$

$$f_r = \frac{1}{2\pi \sqrt{LC}} \qquad \text{(Eq. 5-9)}$$

$$f_r = \frac{1}{2\pi \sqrt{(6.28 \times 10^{-6})(20 \times 10^{-12})}}$$

$$f_r = \frac{1}{2\pi \sqrt{(125.6 \times 10^{-18})}}$$

$$f_r = \frac{1}{(6.28)(11.2 \times 10^{-9})}$$

$$f_r = 14.2 \times 10^6 \ Hz = 14.2 \ MHz$$

Parallel-Resonant Circuits

With a *parallel-resonant circuit* there are several current paths, but the same voltage is applied to the components. Fig. 5-14 shows a parallel LC circuit connected to a signal generator. The series resistor is just a precaution to prevent the circuit from overloading the generator. The applied voltage will force some current through the branches. The current through the coil will be 180° out of phase with the current in the capacitor, and again they add up to zero. So the total current into or out of the generator is very small. It is a mistake to assume, however, that because the generator current is small there is a small current flowing through the capacitor and inductor. In a parallel resonant circuit, the current through the inductor and capacitor is very large. We call this current the circulating current, or tank current.

If we use points A and B in Fig. 5-14A as reference points and examine the branch currents, we observe an interesting phenomenon. Because the current through the capacitor and inductor are 180° out of phase, the branch currents appear to flow in opposite directions. For example, in the figure the current through the inductor may appear to flow in the direction from A to B, while the current through the capacitor is flowing from B to A. When the applied current from the generator reverses polarity, the branch currents reverse direction. From the point of view of the circulating current, the two reactive components are actually in series. Current flows from point A through the inductor to point B, and from point B through the capacitor to point A. At resonance, this circulating current will only be limited by resistive losses in the components and in the wire connecting them.

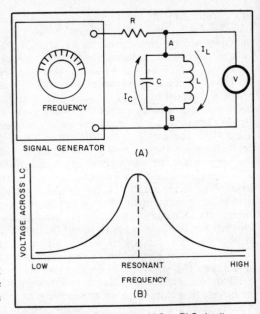

Fig. 5-14 — A parallel-connected LC or RLC circuit behaves like a very high resistance at the resonant frequency. Therefore, the voltage measured across the circuit reaches a peak at the resonant frequency.

We can again think about the energy in the circuit being handed back and forth between the magnetic field of the inductor and the electric field of the capacitor. Large amounts of energy are being transferred, but the generator only has to supply a small amount to make up for the losses in imperfect components. While the total current from the generator is small at resonance, the voltage measured across the tank reaches a maximum value at resonance.

Calculating the resonant frequency of a parallel circuit is exactly the same as for a series circuit. So you should have no trouble finding the resonant frequency of the circuit shown in Fig. 5-14 if R = 1 kΩ, L = 0.12 mH and C = 16 pF. (The answer is 3.6 MHz.)

[Turn to Chapter 10 and use the FCC examination questions with numbers that begin 4AE-5.1 through 4AE-5.20. Review this section as needed.]

Calculating Component Values for Resonance

Using the same technique that we used to derive Eq. 5-9, we can easily derive equations to calculate either the inductance or capacitance to resonate with a certain component at a specific frequency:

$$L = \frac{1}{(2\pi f_r)^2 \, C} \qquad \text{(Eq. 5-10)}$$

$$C = \frac{1}{(2\pi f_r)^2 \, L} \qquad \text{(Eq. 5-11)}$$

You should always be sure to change the frequency to hertz, the capacitance to farads and the inductance to henrys when using these equations.

Let's try solving a couple of problems, just to be sure you know how to handle the equations. Suppose you have a 200-pF capacitor, and you want a circuit that is resonant in the 40-meter band. What inductor value should you choose? Well, first of all, you will have to select a frequency in the 40-meter band to use for the calculation. In a practical design problem, you would have a specific frequency in mind, but for a general example, let's pick some frequency in the middle of the band, such as 7200 kHz. Actually, the required inductance will not change drastically from one end of the band to another. You may want to solve the problem for a resonant frequency of 7000 kHz and 7300 kHz just to prove that to yourself.

Eq. 5-10 will give us the desired answer, but first we must change the frequency to 7200×10^3, or 7.2×10^6 Hz and the capacitance to 200×10^{-12} or 2×10^{-10} F. Then:

$$L = \frac{1}{(2\pi)^2 (7.2 \times 10^6)^2 (2 \times 10^{-10})}$$

$$L = \frac{1}{(39.48) (5.18 \times 10^{13}) (2 \times 10^{-10})}$$

$$L = \frac{1}{4.09 \times 10^5} = 2.44 \times 10^{-6} \text{ H}$$

$$L = 2.44 \ \mu\text{H}$$

So you will need a 2.44-μH inductor to build the desired circuit. If you want a circuit that is resonant at 7.0 MHz, you will need a 2.58-μH inductor, and to make it resonant at 7.3 MHz it will take a 2.38-μH coil.

What value capacitor is needed to make a circuit that is resonant in the 80-meter band if you have a 15-μH coil? Again, choose a frequency in the 80-meter band to work with. Let's pick 3.7 MHz. Then convert to fundamental units: $f_r = 3.7 \times 10^6$ Hz and $L = 15 \times 10^{-6}$ or 1.5×10^{-5} H. This time, select Eq. 5-11, since that one is written to find capacitance, the quantity we are looking for:

$$C = \frac{1}{(2\pi)^2 (3.7 \times 10^6)^2 (1.5 \times 10^{-5})}$$

$$C = \frac{1}{(39.48) (13.69 \times 10^{12}) (1.5 \times 10^{-5})}$$

$$C = \frac{1}{8.11 \times 10^9} = 1.23 \times 10^{-10} \text{ F} = 123 \times 10^{-12} \text{ F}$$

$$C = 123 \text{ pF}$$

If you try solving this problem for both ends of the 80-meter band, you will find that you need a 138-pF capacitor at 3.5 MHz and a 106-pF unit at 4.0 MHz. So any capacitor value within this range will resonate in the 80-meter band with the 15-μH inductor.

[For more practice with this type of problem, study FCC examination questions with numbers that begin 4AG-13. Also study questions with numbers that begin 4AE-2. Review this section as needed.]

Q — THE QUALITY FACTOR OF REAL COMPONENTS

We have talked about ideal resistors, capacitors and inductors, and how they behave in ac circuits. We have shown that resistance in a circuit causes some departure from the ideal conditions. But how can we determine how close to the ideal a certain component comes? Or how much of an effect it will have on the designed circuit conditions? We can assign a number to the coil or capacitor that will tell us

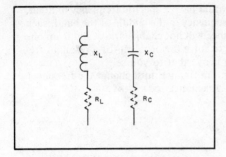

Fig. 5-15 — A practical coil can be considered as an ideal inductor in series with a resistor, and a practical capacitor can be considered as an ideal capacitor in series with a resistor.

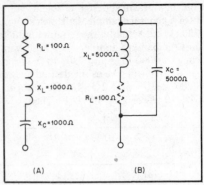

Fig. 5-16 — The Q in a series-resonant circuit such as is shown at A and a parallel-resonant circuit such as is shown at B is found by dividing the inductive reactance by the resistance.

the relative merits of that component — a quality factor of sorts. We call that number Q. We can also assign a Q value to an entire circuit, and that is a measure of how close to the ideal that circuit performs — at least in terms of its resonance properties.

One definition of Q is that it is the ratio of reactance to resistance. Fig. 5-15 shows that a capacitor can be thought of as an ideal capacitor in series with a resistor and a coil can be considered as an ideal inductor in series with a resistor. This internal resistance can't actually be separated from the coil or capacitor, of course, but it acts just the same as if it were in series with an ideal, lossless component. The Q of a real inductor, L, is equal to the inductive reactance divided by the resistance and the Q of a real capacitor, C, is equal to the capacitive reactance divided by the resistance:

$$Q = \frac{X}{R} \qquad\qquad\qquad\qquad\text{(Eq. 5-12)}$$

If you want to know the Q of a circuit containing both internal and external resistance, both resistances must be added together to find the value of R used in the equation. Since added external resistance can only raise the total resistance, the Q always goes down when resistance is added in series. There is no way to lower the internal resistance of a coil or capacitor and raise the Q, therefore, except by building a better component. The internal resistance of a capacitor is usually much less than that for a coil, so we often ignore the resistance of a capacitor and consider only that associated with the coil. Fig. 5-16A shows a series RLC circuit with a Q of 10. To calculate that value, select either value of reactance and divide it by the resistor value.

At Fig. 5-16B, we have increased the input frequency 5 times. The reactance of our inductor has increased 5 times, and we have selected a new capacitor to provide a resonant circuit. This time the components are arranged to provide a parallel-resonant circuit. The circuit Q is still found using Eq. 5-12. You will notice that the Q for this circuit is 50. Increasing the frequency increased the inductive reactance, so as long as the internal resistance stays the same, the Q increases by the same factor.

Skin Effect Increases Resistance

Unfortunately, the internal resistance of the coil (due mainly to the resistance of the wire used to wind it) increases somewhat as the frequency increases. In fact,

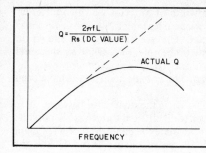

Fig. 5-17 — For low frequencies, the Q of an inductor is proportional to frequency. At high frequencies, increased losses in the coil cause the Q to be degraded from the expected value.

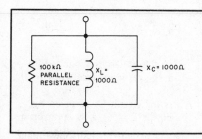

Fig. 5-18 — The Q in a parallel resonant circuit with parallel resistance is found by dividing the parallel-resistance value by the inductive reactance.

the coil Q will increase with increasing frequency up to a point, but then the internal resistance becomes greater and the coil Q degrades. Fig. 5-17 illustrates how coil Q changes with increasing frequency.

The major cause of this increased resistance at higher frequencies is something known as *skin effect*. As the frequency increases, the electric and magnetic fields of the signal do not penetrate as deeply into the conductor. At dc, the entire thickness of the wire is used to carry current, but as the frequency increases, the effective area gets smaller and smaller. In the HF range, the current all flows in the outer few thousandths of an inch of the conductor, and at VHF and UHF, the depth is on the range of a few ten thousandths of an inch. This makes the wire less able to carry the electron flow, and increases the effective resistance.

Q in Parallel-Resonant Circuits

Often a parallel resistor is added to a parallel-resonant circuit to decrease the Q and increase the *bandwidth* of the circuit, as shown in Fig. 5-18. Such a resistor should have a value more than 10 times the reactance of the coil (or capacitor) at resonance or the resonance conditions will change. With the added parallel resistor, circuit Q is found by dividing the resistor value by the reactance value. So for a parallel-resonant circuit:

$$Q = \frac{R}{X} \qquad \text{(Eq. 5-13)}$$

The Q of the circuit shown in Fig. 5-18 is 100.

As in the case of a series circuit, any added resistance degrades the circuit Q and increases the bandwidth. When a resistor is added in parallel, its value is always less than if no resistor were included, because in that case the parallel resistance is infinitely large. Adding a resistor to the circuit always decreases that value. So the larger the value of added resistance, the higher the Q.

Resonant-Circuit Bandwidth

The higher the circuit Q, the sharper the frequency response of a resonant circuit will be. Fig. 5-19 shows the relative bandwidth of a circuit with two different Q values. Bandwidth refers to the frequency range over which the circuit response is no more than 3 dB below the peak response. The −3-dB points are shown on Fig. 5-19, and the bandwidths are indicated. Since this 3-dB decrease in signal represents the points where the circuit power is one half of the resonant power, the −3-dB points are also called *half-power points*. The voltage and current have been reduced to 0.707 times their peak values.

The half-power points are called f_1 and f_2; Δf is the difference between these two frequencies, and represents the half-power (or 3 dB) bandwidth. It is possible to calculate the bandwidth of a resonant circuit based on the circuit Q and the resonant frequency.

$$\Delta f = \frac{f_r}{Q} \qquad \text{(Eq. 5-14)}$$

where

$\quad \Delta f$ is the half-power bandwidth
$\quad f_r$ is the resonant frequency of the circuit
$\quad Q$ is the circuit Q, as given by Eq. 5-12 or 5-13 as appropriate.

Let's calculate the half-power bandwidth of a parallel-resonant circuit such as is shown in Fig. 5-18. The resonant frequency is 7200 kHz; the inductance value is 400 μH; and the parallel resistance is 2 MΩ. Inductive reactance is easily found:

$$X_L = 2\pi fL$$

$$X_L = (6.28)(7.2 \times 10^6 \text{ Hz})(400 \times 10^{-6} \text{ H})$$

$$X_L = 1.81 \times 10^4 \ \Omega = 18.1 \text{ k}\Omega$$

Then the Q is given by Eq. 5-13:

$$Q = \frac{R}{X} = \frac{2 \times 10^6}{18.1 \times 10^3} = 110$$

and the half-power bandwidth is found by Eq. 5-14:

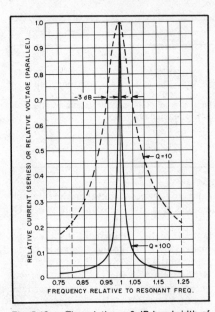

Fig. 5-19 — The relative −3-dB bandwidth of two resonant circuits is shown. The circuit with the higher Q has a steeper response, and a narrower bandwidth. Notice that the vertical scale represents current for a series circuit and voltage for a parallel one. The two circuits are considered to have equal peak response for either case.

$$\Delta f = \frac{f_r}{Q} = \frac{7.2 \times 10^6 \text{ Hz}}{110} = 6.55 \times 10^4 \text{ Hz}$$

$$\Delta f = 65.5 \times 10^3 \text{ Hz} = 65.5 \text{ kHz}$$

Notice that the response of this circuit will be at least half of the peak signal power for signals in the range 7167 to 7233 kHz. To find the upper and lower frequency limits you have to subtract half the total bandwidth from the center frequency for the lower limit and add half the bandwidth to get the upper frequency limit.

[Now study FCC examination questions with numbers that begin 4AE-3 and 4AE-5.21 through 4AE-5.40. Review this section as needed.]

POWER IN REACTIVE CIRCUITS

Earlier in this chapter, we learned that energy is stored in the magnetic field of an inductor when current increases through it, and in the electric field of a capacitor when the voltage across it increases. That energy is returned to the circuit when the current through the inductor decreases or when the voltage across the capacitor decreases. We also learned that the voltages across and currents through these components are 90° out of phase with each other. One way to think of this situation is that in one half of the cycle the power source gives some energy to the inductor or capacitor, only to have the same amount of energy handed back on the next half cycle. A perfect capacitor or coil does not consume any energy, but current does flow in the circuit when a voltage is applied to it.

Power is the time rate of doing work, or the time rate of using energy. Going back to our example of work and energy at the beginning of this chapter, if you did the 10 foot-pounds of work in 5 seconds, then you have developed a power of 2 foot-pounds per second. If you could develop 550 foot-pounds per second of power, then we would say you developed 1 horsepower. So power is a way to express not only how much work you are doing (or how much energy is being stored); it also tells how fast you are doing it. In the metric system of measure, which is used to express all of our common electrical units, power is expressed in terms of the watt, which means energy is being stored at the rate of 1 joule per second, or work is being done at the rate of 1 joule per second. To pass the FCC Element 3 exam for a Technician or General class license, you learned that electrical power is equal to the current times the voltage:

$$P = I E \qquad \text{(Eq. 5-15)}$$

But there is one catch. That equation is only true when the current is through a resistor and the voltage is across that resistor. In other words, the current and voltage must be in phase.

Peak and Average Power

In a dc circuit, it is easy to calculate the power. Simply measure or calculate the current and voltage, and multiply the values. But in an ac circuit, you must specify whether you are using the peak or effective (root-mean-square — RMS) values. Peak voltage times peak current will give the *peak power*, and RMS voltage times RMS current gives the *average-power* value. Notice that we do not call this RMS power! It is interesting to note that with pure sine waves, the peak power is just two times the average power. With a modulated RF wave, which does not result in a pure sine wave, the relationship between peak and average power is not quite that simple. To avoid any possible confusion, we usually refer to the average power output of an RF amplifier. So when you say the output power from your transmitter is 100 W, you mean that the average output power is 100 W.

Another term that is often used in relation to RF power amplifiers is *peak-envelope power (PEP)*. This is the power specification used by the FCC to determine the maximum permissible output power. PEP is defined as the average power of the RF envelope during a modulation peak. So PEP refers to the maximum average power of a modulated wave, and not the peak power as defined in the previous paragraph! It is interesting to note that in the case of a CW transmitter with the key held down, the PEP and average power will be the same, and can be measured on a peak-reading wattmeter.

Power Factor

An ammeter and a voltmeter connected in an ac circuit to measure voltage across and current through an inductor or capacitor will read the correct RMS values, but multiplying them together does not give a true indication of the power being dissipated in the component. If you multiply the RMS values of voltage and current read from these meters, you will get a quantity that is referred to as *apparent power*. This term should tell you that the power found in that way is not quite correct! The apparent power should be expressed in units of volt-amperes (VA) rather than watts. The apparent power in an inductor or capacitor is called *reactive power* or nonproductive, wattless power. Reactive power is expressed in volt-amperes-reactive (VARs).

When there are inductors and capacitors in the circuit, the voltage is not in phase with the current. And, as we have already said, there is no power developed in a circuit containing a pure capacitor or a pure inductor. There is something about this phase angle between the voltage and current that must be taken into account. That something is called the *power factor*. Power factor is a quantity that relates the apparent power in a circuit to the real power. You can find the *real power* in a circuit by:

$$P = I^2 R \qquad\qquad (Eq.\ 5\text{-}16)$$

for a series circuit, or:

$$P = \frac{E^2}{R} \qquad\qquad (Eq.\ 5\text{-}17)$$

for a parallel ciruit. Notice that both of these equations are easily derived by using Ohm's Law to solve for either voltage or current, ($E = I \times R$ and $I = E / R$) and replacing that term with the Ohm's Law equivalent.

One way to calculate the power factor is to simply divide the real power by the apparent power:

$$\text{Power factor} = \frac{P_{REAL}}{P_{APPARENT}} \qquad\qquad (Eq.\ 5\text{-}18)$$

Fig. 5-20 shows a series circuit containing a 75-Ω resistor and a coil with an inductive reactance of 100 Ω at the signal frequency. The voltmeter reads 250-V RMS and the ammeter indicates a current of 2-A RMS. This is an apparent power of 500 VA. Use Eq. 5-16 to calculate the power dissipated in the resistor, $P_{REAL} = (2\ A)^2 \times 75\ \Omega$ $= 4\ A^2 \times 75\ \Omega = 300\ W$. Now by using Eq. 5-18, we can calculate the power factor:

$$\text{Power factor} = \frac{300\ W}{500\ VA} = 0.6$$

Another way to calculate the real power, if you know the power factor, is given by:

$$P_{REAL} = P_{APPARENT} \times \text{Power factor} \qquad\qquad (Eq.\ 5\text{-}19)$$

In our example,

$$P_{REAL} = 500\ VA \times 0.6 = 300\ W$$

Of course the value found using Eq. 5-19 must agree with the value found by either Eq. 5-16 or 5-17, depending on whether the circuit is a series or a parallel one.

What if you don't have the benefit of a voltmeter and an ammeter in a circuit? How can you calculate the real power or power factor? Well, it turns out that the phase angle between the total applied voltage and the circuit current can be used. We already learned how to calculate the phase angle of either a series or a parallel circuit, so that should be no problem. If you don't remember how to calculate the phase angle, go back and review that section of this chapter.

The power factor can be calculated from the phase angle by finding the cosine value of the phase angle:

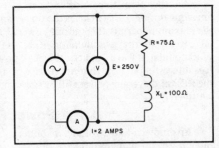

Fig. 5-20 — Only the resistance actually consumes power. The voltmeter and ammeter read the proper RMS values for the circuit, but their product is apparent power, not real average power.

Power factor $= \cos \theta$ (Eq. 5-20)

where θ is the phase angle between voltage and current in the circuit and cosine is a trigonometric function with values that vary between 0 for an angle of $-90°$ (such as for the voltage across a capacitor lagging the current) to 1 for an angle of $0°$ (such as for the voltage and current being in phase for a resistor) and back to 0 for an angle of $90°$ (such as for the voltage across an inductor leading the current).

From this discussion, we can see that for a circuit containing only resistance, where the voltage and current are in phase, the power factor is 1, and the real power is equal to the apparent power. For a circuit containing only pure capacitance or pure inductance, the power factor is 0, so there is no real power! For most practical circuits, which contain resistance, inductance and capacitance, and the phase angle is some value greater than or less than $0°$, the power factor will be something less than one. In such a circuit, the real power will always be something less than the apparent power. This is an important point to remember.

Let's try a sample problem, just to be sure you understand all this. We will assume you can calculate the phase angle between voltage and current, given the component values and generator frequency. After that, it doesn't matter if the circuit is a series or parallel one. The procedure is the same. The current through a particular circuit leads the voltage applied to it by $30°$. If the applied voltage is 200 V and the total current drawn from the source is 3 A, what is the real power in the circuit?

The power factor is found by using Eq. 5-20:

Power factor $= \cos(30°) = 0.866$

The apparent power is calculated using Eq. 5-15:

$P_{APPARENT} = 3\ A \times 200\ V = 600\ VA$

Real power is then found using Eq. 5-19:

$P_{REAL} = 600\ VA \times 0.866 = 520\ W$

[Study FCC examination questions with numbers that begin 4AE-1 and 4AE-7. Review this section as needed.]

EFFECTIVE RADIATED POWER (ERP)

Knowing the output power from your transmitter is important to ensure that you stay within the limits set by FCC rules for your Amateur Radio Station. Sometimes it is more helpful, in evaluating your total station performance, to know how much power is actually being radiated. In fact, if you are responsible for the actual operation of a repeater station, then it is important for you to know how much power is being radiated. The FCC rules governing Amateur Radio repeaters specify the maximum permissible power radiated from the antenna in terms of the antenna height above average terrain. The higher the repeater antenna, the less *effective radiated power (ERP)* you are allowed from the repeater. This specification is based on the ERP from the repeater antenna.

You may be wondering why the transmitter power output is not the same as the power radiated from the antenna. Well, there is always some power lost in the feed line, and often there are other devices inserted in the line, such as a wattmeter, SWR bridge or an impedance-matching network. In the case of a repeater system, there is usually a duplexer so the transmitter and receiver can use the same antenna, and perhaps a circulator to reduce the possibility of intermodulation interference. These devices also introduce some loss to the system. Antennas are compared to a reference, and they may exhibit gain or loss as compared to that reference.

The two types of antennas commonly used for reference are a half-wave dipole and a theoretical isotropic radiator. For our discussion and calculations here, we will assume the antenna gain is with reference to a half-wave dipole. You will learn more about the isotropic radiator when you study for the Amateur Extra Class license. A beam antenna will have some gain over a dipole, at least in the desired radiation direction. The exact amount of gain will depend on the design and installation.

These system gains and losses are usually expressed in *decibels*. The decibel is a logarithm of the ratio of two power levels, and the gain in dB is calculated by:

$$dB = 10 \log \left(\frac{P2}{P1} \right) \qquad \text{(Eq. 5-21)}$$

where
  P1 is the reference power
  P2 is the power being compared to the reference

In the case of calculating ERP, the transmitter output power is considered as the reference, and the power at any other point is P2.

The main advantage of using decibels is that system gains and losses expressed in these units can simply be added, with losses written as negative values. If we are using a 150-W transmitter followed by 6 dB of circulator and duplexer loss, 3 dB of feed line loss and a 4-dB gain antenna, our total system gain looks like:

System gain = -6 dB + -3 dB + 4 dB = -5 dB

Note that this is a **loss** of 5 dB total for the system. Using Eq. 5-21:

$$-5 \text{ dB} = 10 \log \left(\frac{P2}{150 \text{ W}} \right)$$

$$\log^{-1} \left(\frac{-5 \text{ dB}}{10} \right) = \log^{-1} \left(\log \left(\frac{P2}{150 \text{ W}} \right) \right)$$

(log $^{-1}$ means the antilog, or inverse log function)

$$\log^{-1} (-0.5 \text{ dB}) = \frac{P2}{150 \text{ W}}$$

$$P2 = \log^{-1} (-0.5) \times 150 \text{ W} = 0.32 \times 150 \text{ W} = 48 \text{ W}$$

This is consistent with our expectation that with a 5-dB system loss we would have less ERP than transmitter output power.

Suppose we have a repeater transmitter that has a 200-W output, followed by a duplexer and a circulator that have 1.5-dB loss each and a feed line that exhibits 3 dB loss at the output frequency. The signal is then fed to a 12-dB gain antenna. What is the effective radiated power from the antenna? To calculate the total system gain (or loss) we add the decibel values given:

System gain = -1.5 dB + -1.5 dB + -3 dB + 12 dB
System gain = -6 dB + 12 dB = 6 dB

Then we can use Eq. 5-21 to find the ERP:

$$6 \text{ dB} = 10 \log \left(\frac{P2}{200 \text{ W}} \right)$$

$$\log^{-1} \left(\frac{6 \text{ dB}}{10} \right) = \log^{-1} \left(\log \left(\frac{P2}{200 \text{ W}} \right) \right)$$

$$\log^{-1} (0.6 \text{ dB}) = \frac{P2}{200 \text{ W}}$$

$$P2 = \log^{-1} (0.6) \times 200 \text{ W} = 4.0 \times 200 \text{ W} = 800 \text{ W}$$

As long as the antenna for your 2-meter repeater is no more than 105 feet above the average surrounding terrain, this system will satisfy the FCC power limitation. For more information about maximum ERP for repeater systems see the ARRL publication *The FCC Rule Book*. *The ARRL Handbook* contains information about calculating height above average terrain, if you are interested. You do not have to know how to calculate height above average terrain for the Advanced class license, however.

[Study FCC examination questions with numbers that begin 4AE-8 at this point. If you have any difficulty with those questions, review this section.]

THEVENIN'S THEOREM

Thevenin's Theorem is a useful tool for simplifying complex networks of resistors (or other components). Our discussion here is limited to resistors, to make it easier for you to understand. (Besides, you don't have to know how to handle this technique with reactive components for the Element 4A examination!) The theorem states that any two-terminal network of resistors and voltage sources, no matter how complex, can be replaced by a circuit consisting of a single voltage source and a single series resistor. Doing this conversion for a part of a complex circuit greatly simplifies calculations involving other parts of the circuit.

We will use a simple example to illustrate the technique provided by Thevenin's Theorem. While it may seem easier to solve this problem using other methods that you are familiar with, the advantages become obvious with more complex problems. Fig. 5-21 shows a simple circuit involving a voltage source, two parallel resistors and a series resistor. We want to know the current through R3, so that is considered as the load for the Thevenin-equivalent circuit.

We could solve this problem by calculating the resistance of the parallel-resistor combination, adding the series resistence, then using Ohm's Law to find the total circuit current. By calculating the voltage drops across the series resistor and the parallel pair, we could finally solve for the current through R3. A lot of work, even for this simple problem.

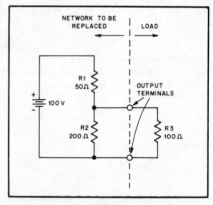

Fig. 5-21 — Circuit showing a combination of a voltage source and a resistive network that can be simplified using Thevenin's Theorem to find the load current.

To apply Thevenin's Theorem, first remove the load from the network to be simplified. The new Thevenin-equivalent voltage source has a voltage equal to the open-circuit (no load) voltage at the output terminals. Fig. 5-22A shows the network to be replaced, with the output voltage. You can find that output voltage by using Ohm's Law to calculate the total circuit current and the voltage drop across R2:

$$V_{OUT} = I \times R2$$

But the circuit current is equal to the source voltage divided by the total resistance:

$$I = \frac{E}{R1 + R2}$$

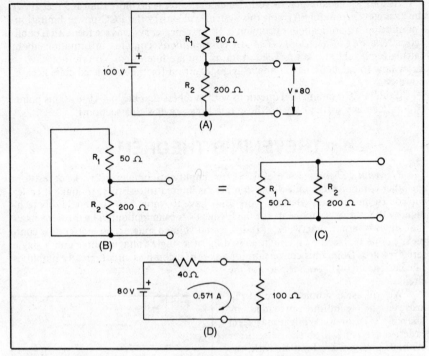

Fig. 5-22 — Steps showing how a two-terminal network can be simplified to find a Thevenin-equivalent voltage source (A) and the Thevenin-equivalent resistance (B and C) to calculate the load current (D).

Then:

$$V_{OUT} = \frac{E \times R2}{R1 + R2} = \frac{100 \text{ V} \times 200 \text{ }\Omega}{50 \text{ }\Omega + 200 \text{ }\Omega} = \frac{20000 \text{ V}\Omega}{250 \text{ }\Omega} = 80 \text{ V}$$

Now replace the voltage source in the original network with its "internal resistance." (Most voltage sources can be assumed to have an internal resistance of zero ohms, so we will use a piece of wire in place of the voltage source.) See Fig. 5-22B. The Thevenin-equivalent series resistor is equal to that resistance which would be measured by an ohmmeter connected to the load terminals of the circuit with the voltage source shorted out. We can redraw the circuit of Fig. 5-22B as shown in part C, which clearly shows that these two resistors are simply in parallel. Resistors in parallel are combined by adding their reciprocals, then taking the reciprocal of that sum. When there are only two resistors to combine, you probably remember the equation from when you studied for the Technician/General class exam:

$$R_T = \frac{R1 \times R2}{R1 + R2} \qquad \text{(Eq. 5-22)}$$

$$R_T = \frac{50 \text{ }\Omega \times 200 \text{ }\Omega}{50 \text{ }\Omega + 200 \text{ }\Omega} = \frac{10000 \text{ }\Omega^2}{250 \text{ }\Omega}$$

$$R_T = 40 \text{ }\Omega$$

Now we can replace the original voltage-divider network with our new Thevenin-equivalent circuit, consisting of an 80-V source in series with a 40-ohm resistor. Adding the load resistor back in, as shown in Fig. 5-22D, enables us to calculate the load current easily using Ohm's Law:

$$I = \frac{E}{R_T + R_3} = \frac{80 \text{ V}}{40 \text{ } \Omega + 100 \text{ } \Omega} = \frac{80 \text{ V}}{140 \text{ } \Omega}$$

$$I = 0.57 \text{ A}$$

The whole procedure is much more difficult to describe than it is to carry out, and after practicing a few problems, you should be ready to tackle problems with much more complicated networks, and maybe even a few inductors and capacitors thrown in with an ac voltage source!

[To get some of that practice that you need, turn to Chapter 10, and try the FCC examination questions with numbers that begin 4AE-9. Review this section as needed.]

Key Words

Alpha (α) — The ratio of transistor collector current to emitter current. It is between 0.92 and 0.98 for a junction transistor.

Alpha cutoff frequency — A term used to express the useful upper frequency limit of a transistor. The point at which the gain of a common-base amplifier is 0.707 times the gain at 1 kHz.

Anode — The terminal that connects to the positive supply lead for current to flow through a device.

Avalanche point — That point on a diode characteristic curve where the amount of reverse current increases greatly for small increases in reverse bias voltage.

Beta (β) — The ratio of transistor collector current to base current. Betas show wide variations, even between individual devices of the same type.

Beta cutoff frequency — The point at which the gain of a common-emitter amplifier is 0.707 times the gain at 1 kHz.

Bipolar junction transistor — A transistor made of two PN semiconductor junctions using two layers of similar-type material (N or P) with a third layer of the opposite type between them.

Cathode — The terminal that connects to the negative supply lead for current to flow through the device.

Crystal-lattice filter — A filter that employs piezoelectric crystals (usually quartz) as the reactive elements. They are most often used in the IF stages of a receiver or transmitter.

Depletion region — An area around the semiconductor junction where the charge density is very small. This creates a potential barrier for current flow across the junction. In general, the region is thin when the junction is forward biased, and becomes thicker under reverse-bias conditions.

Doping — The addition of impurities to a semiconductor material, with the intent to provide either excess electrons or positive charge carriers (holes) in the material.

Forward bias — A voltage applied across a semiconductor junction so that it will tend to produce current flow.

Hot-carrier diode — A type of diode in which a small metal dot is placed on a single semiconductor layer. It is superior to a point-contact diode in most respects.

Light-emitting diode — A device that uses a semiconductor junction to produce light when current flows through it.

Neon lamp — A cold-cathode (no heater or filament), gas-filled tube used to give a visual indication of voltage in a circuit, or of an RF field.

N-type material — A semiconductor material that has been "doped" with excess electrons, so that it becomes a donor material.

Peak inverse voltage — The maximum instantaneous anode-to-cathode reverse voltage that is to be applied to a diode.

PIN diode — A diode consisting of a relatively thick layer of nearly pure semiconductor material with a layer of P-type material on one side and a layer of N-type material on the other.

PN-junction — The contact area between two layers of opposite-type semiconductor material.

Point-contact diode — A diode that is made by a pressure contact between a semiconductor material and a metal point.

P-type material — A semiconductor material that has been "doped" with excess positive charge carriers or "holes", so that it becomes an acceptor material.

Reverse bias — A voltage applied across a semiconductor junction so that it will tend to prevent current flow.

Semiconductor material — A material with resistivity between that of metals and insulators. Pure semiconductor materials are usually doped with impurities to control the electrical properties.

Silicon-controlled rectifier — A bistable semiconductor device that can be switched between the off and on states by a control voltage.

Thyristor — Another name for a silicon-controlled rectifier.

Triac — A bidirectional SCR, primarily used to control ac voltages.

Tunnel diode — A diode with an especially thin depletion region, so that it exhibits a negative resistance characteristic.

Unijunction transistor — A three-terminal, single-junction device that exhibits negative resistance and switching characteristics unlike bipolar transistors.

Varactor — A diode that has a voltage-variable capacitance.

Zener diode — A diode that is designed to be operated in the reverse-breakdown region of its characteristic curve.

Zener voltage — A reverse-bias voltage that produces a sudden change in apparent resistance across the diode junction, from a large value to a small value.

Chapter 6

Circuit Components

B efore you can understand the operation of most complex electronic circuits, you must know some basic information about the parts that make up those circuits. This chapter presents the information about circuit components that you need to know in order to pass your Advanced class Amateur Radio license exam. You will find descriptions of several types of diodes, a variety of transistors, silicon-controlled rectifiers, light-emitting diodes, neon lamps and crystal-lattice filters. These components are combined with other devices to build practical electronic circuits, which are described in the next chapter.

As you study the characteristics of the components described in this chapter, be sure to turn to the FCC Element 4A questions in Chapter 10 when you are directed to do so. That will provide a review of bite-sized chunks of the material, as it is covered. It will show you where you need to do some extra studying, and maybe even where you need to turn to some other sources of material. If you thoroughly understand how these components work, then you should have no problem learning how they can be connected to make a circuit perform a specific task.

DIODES

To qualify for the Advanced class license, you must be familiar with some specialized types of diodes. Rectifier circuits were covered on the Technician/General examination; for the Advanced class test, you must be familiar with the two main structural categories of diodes: junction and point-contact types. PIN, Zener, tunnel, varactor and hot-carrier diodes are special-purpose devices that come under those main headings.

Junction Diodes

Semiconductor material exhibits properties of both metallic and nonmetallic substances. Generally, the layers of semiconductor material are made from germanium or silicon. To control the conductive properties of the material, impurities are added as the semiconductor crystal is "grown." This process is called *doping*. If a material with an excess of free electrons is added to the structure, *N-type material* is produced. This type of material is sometimes referred to as donor material. When an impurity that results in an absence of electrons is introduced to the crystal structure, you get *P-type material*. The result in this case is a material with positive charge carriers, called holes. P-type material is also called acceptor material.

The junction diode, also called the *PN-junction* diode, is made from two layers of semiconductor material joined together. One layer is made from P-type (positive) material. The other layer is made from N-type (negative) material. The name PN junction comes from the way the P and N layers are joined to form a semiconductor diode. Fig. 6-1 illustrates the basic concept of a junction diode.

The P-type side of the diode is called the *anode*, while the N-type side is called the *cathode*. Current flow in a diode is from the cathode to the anode; that is, the excess electrons from the N-type material flow to the P-type material, which has holes (electron deficiency). Electrons and holes are also called carriers because they are the means by which current is carried from one side of the junction to the other. When no voltage is applied to a diode, the junction between the P-type and N-type material acts as a barrier that prevents carriers from flowing between the layers.

When voltage is applied to a junction diode as shown at A in Fig. 6-2, however, carriers will flow across the barrier and the diode will conduct (that is, electrons will flow through it). When the diode anode is positive with respect to the cathode, electrons are attracted across the junction from the N-type material, through the P-type material and on to the positive battery terminal. Holes are attracted in the opposite direction by the negative potential of the battery. When the diode anode is connected in this manner it is said to be *forward biased*.

If battery polarity is reversed, as shown at B in Fig. 6-2, the excess electrons in the N-type material are attracted away from the junction by the positive battery terminal. Similarly, the holes in the P-type material are attracted away from the junction by the negative battery side. When this happens, the area around the junction has no current carriers; electrons do not flow across the junction to the P-type material, and the diode does not conduct. When the anode is connected to a negative voltage source and the cathode is connected to a positive voltage source, the diode does not conduct, and the device is said to be *reverse biased*.

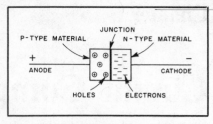

Fig. 6-1 — A PN junction consists of P-type and N-type material separated by a barrier.

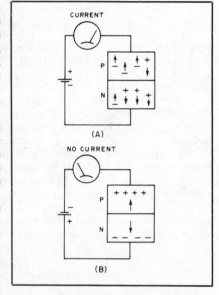

Fig. 6-2 — At A, the PN junction is forward biased and conducting; at B, it is reverse biased, so it does not conduct.

Junction diodes are used as rectifiers because they allow current flow in one direction only. See Fig. 6-3. When an ac signal is applied to a diode, it will be forward biased during one half of the cycle, so it will conduct, and current will flow to the load. During the other half of the cycle, the diode is reverse biased, and no current will flow. The diode output is pulsed dc, and current always flows in the same direction.

In a junction diode, the P and N layers are sandwiched together, separated by the junction. Although the spacing between the layers is extremely small, there is some capacitance at the junction. The structure can be thought of in much the same way as a simple capacitor: two charged plates separated by a thin dielectric.

Although the internal capacitance of a PN-junction diode may be only a few

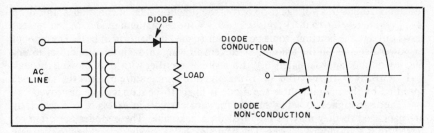

Fig. 6-3 — PN-junction diodes are used as rectifiers because they allow current flow in one direction only.

picofarads, this capacitance can cause problems in RF circuits, especially at VHF and above. Junction diodes may be used from dc to the microwave region, but there is a special type of diode with low internal capacitance that is specially designed for RF applications. This device, called the *point-contact diode*, is discussed later in this chapter.

Diode Ratings

Junction diodes have maximum voltage and current ratings that must be observed, or damage to the diode could result. The voltage rating is called *peak inverse voltage* (PIV), and the current rating is called average rectified forward current. With present technology, diodes are commonly available with ratings up to 1000 PIV and 100 A.

Peak inverse voltage is the voltage that a diode must withstand when it isn't conducting. Although a diode is normally used in the forward direction, it will conduct in the reverse direction if enough voltage is applied. A few hole/electron pairs are thermally generated at the junction when a diode is reverse biased. These pairs cause a very small reverse current, called leakage current, to flow. Semiconductor diodes can withstand some leakage current. If the inverse voltage reaches a high enough value, however, the leakage current rises abruptly, resulting in a heavy reverse current flow. The point where the leakage current rises abruptly is called the *avalanche point*. A large reverse current flow usually damages or destroys the diode.

The maximum average forward current is the highest average current that can flow through the diode in the forward direction for a specified junction temperature. This specification varies from device to device, and it depends on the maximum allowable junction temperature and on the amount of heat the device can dissipate. As the forward current increases, the junction temperature will increase. If allowed to get too hot, the diode will be destroyed.

Impurities at the PN junction cause some resistance to current flow. This

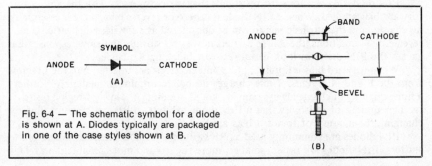

Fig. 6-4 — The schematic symbol for a diode is shown at A. Diodes typically are packaged in one of the case styles shown at B.

resistance results in a voltage drop across the junction. For silicon diodes, this drop is approximately 0.6 to 0.7 V; it is 0.2 to 0.3 V for germanium diodes. When current flows through the junction, some power is dissipated in the form of heat. The amount of power depends on the amount of current flowing through the diode. For example, it would be approximately 6 W for a silicon rectifier with 10 A flowing through it (P = IE; P = 10 A × 0.6 V). If the junction temperature exceeds the safe level specified by the manufacturer the diode is likely to be damaged or destroyed.

Diodes designed to safely handle forward currents in excess of 6 A generally are packaged so they may be mounted on a heat sink. These diodes are often referred to as stud-mount devices. The heat sink helps the diode package to dissipate heat more rapidly, thereby keeping the diode junction temperature at a safe level. The metal case of a stud-mount diode is usually one of the contact points, so it must be insulated from ground.

Fig. 6-4 shows some of the more common diode-case styles, as well as the general schematic symbol for a diode. The line, or spot, on a diode case indicates the cathode lead. On a high-power, stud-mount diode, the stud may be either the anode or cathode. Check the case or the manufacturer's data sheet for the correct polarity.

Varactor and Varicap® Diodes

As mentioned before, junction diodes exhibit an appreciable internal capacitance. It is possible to change the internal capacitance of a diode by varying the amount of reverse bias applied to it. Manufacturers have designed certain kinds of diodes, called voltage-variable capacitors or variable-capacitance diodes (Varicaps) and *varactors* (variable reactance diodes) to take advantage of this property.

Varactors are designed to provide various capacitance ranges from a few picofarads to more than 100 pF. Each style has a specific minimum and maximum capacitance, and the higher the maximum amount, the greater the minimum amount. A typical varactor can provide capacitance changes over a 10:1 range with bias voltages in the 0- to 100-V range.

Varactors are similar in appearance to junction diodes. Common schematic symbols for a varactor diode are given in Fig. 6-5. These devices are used in frequency multipliers at power levels as great as 25 W, remotely tuned circuits and simple frequency modulators.

PIN Diodes

A *PIN* (positive/intrinsic/negative) *diode* is formed by diffusing P-type and N-type layers onto opposite sides of an almost pure silicon layer, called the I region. See Fig. 6-6. This layer is not "doped" with P-type or N-type charge carriers, as are the other layers. Any charge carriers found in this layer are a result of the natural properties of the pure semiconductor material. In the case of silicon, there are relatively few free charge carriers. PIN-diode characteristics are determined primarily by the thickness and area of the I region.

PIN diodes have a forward resistance that varies inversely with the amount of forward bias applied. When a PIN diode is at zero or reverse bias, there is essentially no charge, and the intrinsic region can be considered as a low-loss dielectric. Under reverse-bias conditions, the charge carriers move very slowly. This slow response time causes the PIN diode to look like a resistor, blocking RF currents.

When forward bias is applied, holes and electrons are injected into the I region from the P and N regions. These charges do not recombine immediately. Rather, a finite quantity of charge always remains stored, resulting in a lower I-region resistivity. The amount of resistivity that a PIN diode exhibits to RF can be controlled by changing the amount of forward bias applied.

PIN diodes are commonly used as RF switches, variable attenuators and phase shifters. PIN diodes are faster, smaller, more rugged and more reliable than relays or other electromechanical switching devices.

Fig. 6-7 shows how a PIN diode may be used as an SPST RF switch. With no bias, or with reverse bias applied to the diode, the PIN diode exhibits a high resistance to RF, so no signal will flow from the generator to the load. When forward bias is applied, the diode resistance will decrease, allowing the RF signal to pass. The amount of insertion loss (resistance to RF current) is determined primarily by the amount of forward bias applied; the greater the forward bias current, the lower the RF resistance.

A PIN-diode attenuator is shown in Fig. 6-8. The PIN diodes are connected as resistors would be in a standard Pi-type resistive pad. The major difference, however, is that the attenuation of this pad can be varied by changing the value of forward bias applied to the diodes, changing their RF resistivity.

PIN diodes are packaged in case styles similar to conventional diodes. The package size depends on the intended application. PIN diodes intended for low-power UHF and microwave work are packaged in small epoxy or glass cases to minimize internal capacitance. Others, intended for high-power switching, are of the stud-mount variety so they can be attached to heat sinks. PIN diodes are shown by the schematic symbol shown in Fig. 6-4.

Zener Diodes

Zener diodes are a special class of junction diode used as voltage references and as voltage regulators. When they are used as voltage regulators, Zener diodes provide a nearly constant dc output voltage, even though there may be large changes in load resistance or input voltage. As voltage references, they provide an extremely stable voltage that remains constant over a wide temperature range.

As discussed earlier, leakage current rises as reverse (inverse) voltage is applied to a diode. At first, leakage current is very small and changes very little with increasing reverse voltage. There is a point, however, where the leakage current rises suddenly. Beyond this point, the current

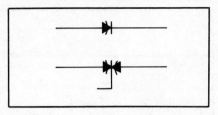
Fig. 6-5 — Schematic symbols for varactor diodes.

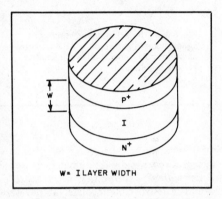
Fig. 6-6 — Inner structure of a PIN diode.

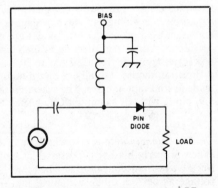
Fig. 6-7 — PIN diodes may be used as RF switches.

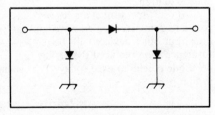
Fig. 6-8 — PIN diodes are often used as variable RF attenuators.

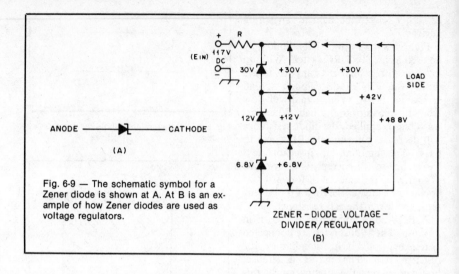

Fig. 6-9 — The schematic symbol for a Zener diode is shown at A. At B is an example of how Zener diodes are used as voltage regulators.

ANODE ——▶|—— CATHODE

(A)

ZENER – DIODE VOLTAGE – DIVIDER/ REGULATOR

(B)

increases very rapidly for a small increase in voltage; this is called the avalanche point. *Zener voltage* is the voltage necessary to cause avalanche. Normal junction diodes would be destroyed immediately if they were operated in this region, but Zener diodes are specially manufactured to withstand the avalanche current safely.

Since the current in the avalanche region can change over a wide range while the voltage stays practically constant, this kind of diode can be used as a voltage regulator. The voltage at which avalanche occurs can be controlled precisely in the manufacturing process. Zener diodes are calibrated in terms of avalanche voltage.

Zener diodes are currently available with ratings between 1.8 and 200 V. The power ratings range from 250 mW to 50 W. They are packaged in the same case styles as junction diodes. Usually, Zener diodes rated for 10-W dissipation or more are made in stud-mount cases. The schematic symbol for a Zener diode is shown in Fig. 6-9, along with an example of how such a device is used as a voltage regulator.

Tunnel Diodes

The *tunnel diode* is a special type of device that has no rectifying properties. When properly biased, it possesses an unusual characteristic: negative resistance. Negative resistance means that when the voltage across the diode increases, the current decreases. This property makes the tunnel diode capable of amplification and oscillation.

At one time, tunnel diodes were expected to dominate in microwave applications, but other devices with better performance soon replaced them. The tunnel diode is seldom used today. The schematic symbol for a tunnel diode is shown in Fig. 6-10.

Fig. 6-10 — Schematic symbol for a tunnel diode.

Point-Contact Diodes

Fig. 6-11 illustrates the internal structural differences between a junction diode and a point-contact diode. As you can see by this diagram, the point-contact diode

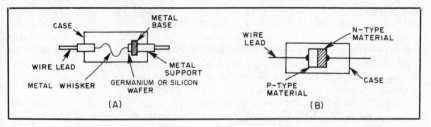

Fig. 6-11 — The internal structure of a point-contact diode is shown at A; at B is the internal structure of a PN-junction diode.

has a much smaller surface area at the junction than does a PN-junction diode. When a point-contact diode is manufactured, the main portion of the device is made from one type of material (P or N), and a tiny region of the other type of material is formed right at the contact point. A thin wire, or metal whisker, connects the contact point to the lead that protrudes through the diode case. The result is a diode that exhibits much less internal capacitance than PN-junction diodes, typically 1 pF or less. This means point-contact diodes are better suited for VHF and UHF work than are PN-junction diodes.

Point-contact diodes are packaged in a variety of cases, as are junction diodes. The schematic symbol is shown in Fig. 6-4. Point-contact diodes are generally used as UHF mixers and as RF detectors at VHF and below.

Hot-Carrier Diodes

Another type of diode with low internal capacitance and good high-frequency characteristics is the *hot-carrier diode*. This device is very similar in construction to the point-contact diode, but with an important difference. Compare the inner structure of the hot-carrier diode depicted in Fig. 6-12 to the point-contact diode shown in Fig. 6-11.

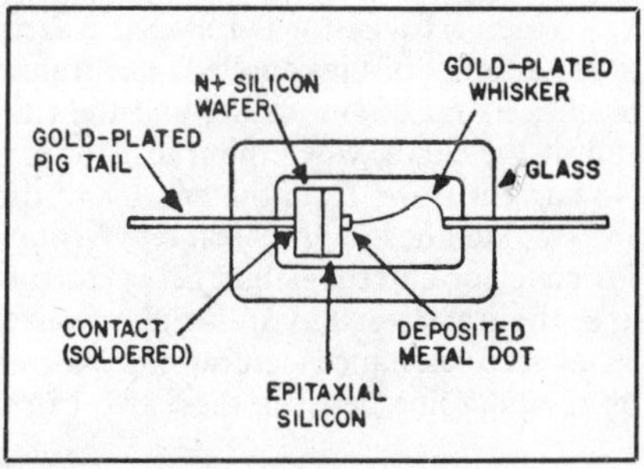

The point-contact device relies on the touch of a metal whisker to make contact with the active element. In contrast, the whisker in a hot-carrier diode is physically attached to a metal dot that is deposited on the element. The hot-carrier diode is

Fig. 6-12 — Internal structure of a hot-carrier diode.

mechanically and electrically superior to the point-contact diode. Some of the advantages of the hot-carrier type are improved power-handling characteristics, lower contact resistance and improved immunity to burnout caused by transient noise pulses.

Hot-carrier diodes are similar in appearance to point-contact and junction diodes and share the same schematic symbol. They are often used in mixers and detectors at VHF and UHF. In this application, hot-carrier diodes are superior to point-contact diodes because they exhibit greater conversion efficiency and lower noise figure.

[Now study FCC examination questions with numbers that begin 4AF-1. Review this section as needed.]

TRANSISTORS

The *bipolar junction transistor* is a type of three-terminal, PN-junction device that is able to amplify signal energy (current). It is made up of two layers of like semiconductor material with a layer of the opposite-type material sandwiched in between. See Fig. 6-13. If the outer layers are P-type material, and the middle layer is N-type material, the device is called a PNP transistor because of the layer arrangement. If the outer layers are N-type material, the device is called an NPN transistor. A transistor is, in effect, two PN-junction diodes back-to-back. Fig. 6-14 depicts the schematic symbols for PNP and NPN bipolar transistors.

The three layers of the transistor sandwich are called the emitter, base and collector. These are functionally analogous to the cathode, grid and plate of a vacuum tube. A diagram of the construction of a typical PNP transistor is given in Fig. 6-13. In an actual bipolar transistor, the center layer (in this case, N-type material) is much thinner than the outer layers. As shown in the diagram, forward-bias voltage across the emitter-base section of the sandwich causes electrons to flow through it from the base to the emitter. As the free electrons from the N-type material flow into the holes of the P-type material, the holes in effect travel into the base. Some of the holes will be neutralized in the base by free electrons, but because the base layer is so thin, some will move right on through into the P-type material in the collector.

As shown, the collector is connected to a negative voltage with respect to the base. Normally, no current will flow because the base-collector junction is reverse biased. The collector, however, now contains an excess of holes because of those from the emitter that overshot the base. Since the voltage source connected to the collector produces a negative charge, the holes from the emitter will be attracted to the power-supply connection. The amount of emitter-to-collector current flow is approximately proportional to the base-to-emitter current. Because of the transistor construction, however, the current flowing through the collector will be considerably larger than that flowing through the base.

When a transistor is forward biased, collector current increases in proportion to the amount of bias applied. The transistor is saturated when the collector current reaches its maximum value, and the transistor is said to be fully on. At the other end of the curve, when the transistor is reverse-biased, the transistor is turned off. No current flows from the emitter to the collector, and the transistor is at cutoff.

A load line is a graphical representation of the range of the transistor resistance for collector-current points between cutoff and saturation. At one end of the load line, the transistor has an infinite resistance to current flow (cutoff); at the other, it has zero resistance (saturation). Normally, a transistor is operated at some point on the load line between these two extremes. For best efficiency and stability, the

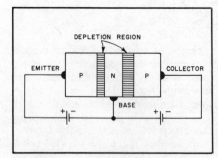

Fig. 6-13 — A bipolar junction transistor consists of two layers of like semiconductor material separated by a layer of the opposite material. This drawing represents the internal structure of a PNP transistor.

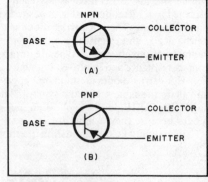

Fig. 6-14 — Schematic symbols for NPN and PNP bipolar transistors.

transistor in a solid-state power amplifier is operated at a point on the load line that is just below saturation.

As shown in Fig. 6-13, there is an area around each junction that is called the *depletion* (sometimes called transition) *region*. The depletion region is an area near a PN junction that is devoid of holes and excess electrons. This region is caused by the repelling forces of the ions on opposite sides of the junction. When the PN junction is reverse biased, the depletion region becomes larger because the electrons and holes are attracted away from the junction. When the PN junction is forward biased, the depletion region becomes smaller because the electrons and holes move toward each other.

Transistor Characteristics

As mentioned before, the current flowing through the collector of a bipolar transistor is approximately proportional to the current flowing through the base. The ratio of collector current to base current is called the current gain, or *beta*. Beta is expressed by the Greek symbol β. It can be calculated from the equation

$$\beta = \frac{I_c}{I_b} \qquad \text{(Eq. 6-1)}$$

where
I$_c$ is the collector current
I$_b$ is the base current

For example, if a 1-mA base current causes a collector current of 100 mA to flow, the beta is 100. Typical betas for junction transistors range from as low as 10 to as high as several hundred. Manufacturers' data sheets specify a range of values for β. Individual transistors of a given type will have widely varying betas.

Another important transistor characteristic is *alpha,* expressed by the Greek letter α. Alpha is the ratio of collector current to emitter current, given by:

$$\alpha = \frac{I_c}{I_e} \qquad \text{(Eq. 6-2)}$$

The smaller the base current, the closer the collector current comes to being equal to that of the emitter, and the closer alpha comes to being 1. For a junction transistor, alpha is usually between 0.92 and 0.98.

Transistors have important frequency characteristics. The *alpha cutoff frequency* is the frequency at which the current gain of a transistor in the common-base configuration decreases to 0.707 times its gain at 1 kHz. Alpha cutoff frequency is considered to be the practical upper frequency limit of a transistor configured as a common-base amplifier.

Beta cutoff frequency is similar to alpha cutoff frequency, but it applies to transistors connected as common-emitter amplifiers. Beta cutoff frequency is the frequency at which the current gain of a transistor in the common-emitter configuration decreases to 0.707 times its gain at 1 kHz.

Bipolar junction transistors are used in a wide variety of applications, including amplifiers (from very low level to very high power), oscillators and power supplies. They are used at all frequency ranges from dc through the UHF and

Fig. 6-15 — Transistors are packaged in a wide variety of case styles, depending on intended application.

microwave range. Transistors are packaged in a wide variety of case styles. Some of the more common case styles are depicted in Fig. 6-15.

Unijunction Transistors

Another three-terminal semiconductor device is the *unijunction transistor* (UJT), sometimes called a double-base diode. The internal structure of a UJT is shown in Fig. 6-16. The elements of a UJT are base 1, base 2 and emitter. There is only one PN junction, and this is between the emitter and the silicon substrate. The base terminals are ohmic contacts; that is, the current is a linear function of the applied voltage. Current flowing between the bases sets up a voltage gradient along the substrate. In operation, the direction of flow causes the emitter junction to be reverse biased.

The most common application for the UJT is in relaxation oscillator circuits. UJTs are packaged in cases similar to small-signal bipolar transistors. Two schematic symbols for UJTs are given in Fig. 6-17. The substrate is made from one type of semiconductor material, and the emitter is made from the other type of material. The large block of material the two bases are connected to in Fig. 6-16 is called the substrate. When the substrate

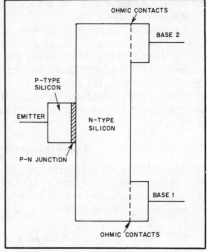

Fig. 6-16 — Internal structure of a unijunction transistor.

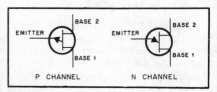

Fig. 6-17 — Schematic symbol of a unijunction transistor.

is N-type material, we call the UJT an N-channel device. Similarly, the substrate of a P-channel UJT is formed from P-type semiconductor material.

[Proceed to Chapter 10 and study FCC examination questions with numbers that begin 4AF-2. Review this section as needed.]

SILICON-CONTROLLED RECTIFIERS

Also known as a *thyristor,* a *silicon-controlled rectifier* (SCR) is a three-terminal, solid-state device. The three terminals are called the cathode, anode and gate. An SCR is a diode whose forward conduction from cathode to anode is controlled by a third terminal, the gate. The schematic symbol for the SCR is shown in Fig. 6-18, along with an equivalent circuit made from discrete components.

In operation, the SCR will not conduct until the voltage across its cathode and anode terminals exceeds the forward-breakdown voltage. Forward-breakdown voltage is determined by the gate current. Without gate current, the SCR presents an open circuit in both directions. When sufficient gate current is applied, the SCR is "triggered" (begins to conduct) and looks like a closed circuit in the forward direction. Once the SCR is triggered, the gate no longer has any control, and the device behaves like a forward-biased junction diode that is conducting. The SCR will continue to conduct, regardless of gate current, until the anode voltage drops to zero again; when this happens, the forward-breakdown voltage barrier is reestablished, and the gate regains control. SCRs have only two stable operating conditions; conducting and nonconducting. There is no partially conducting condition.

SCRs come in a variety of case styles, depending on intended application. See Fig. 6-19. They are often used in power-supply overvoltage-protection circuits (also called crowbars), electronic ignition systems, alarms, and many other applications that require high-speed, unidirectional dc switching.

Triacs

A *triac* is a type of bidirectional SCR. Electrically, a triac is equivalent to two reverse-connected (anode-to-cathode) SCRs wired in parallel, with their gates tied together. Fig. 6-20 shows the schematic symbol for the triac. Its three leads are called anode 1, anode 2 and gate. When the gate of a triac is triggered, the device will conduct either polarity of applied voltage, so triacs are used to switch alternating currents.

The gate voltage of a triac can be set so that it conducts only during part of an ac waveform cycle. Varying the gate voltage varies the amount of the ac voltage that passes through. Triacs can control the current flow to ac-operated devices, and find common application as light dimmers and motor speed controls.

[Study FCC examination questions with numbers that begin 4AF-3. Review this section as needed.]

LIGHT-EMITTING DIODES

Light-emitting diodes (LEDs) are designed to emit light when they are forward biased, and current passes through their PN junctions. The junction of an LED is made from gallium arsenide, gallium phosphide or a combination of these two materials. The color and intensity of the LED depends on the material, or combination of materials, used for the junction. LED colors available today are red, green, orange, white and yellow.

LEDs are packaged in plastic cases, or in metal cases with a transparent end. LEDs are useful as replacements for incandescent panel and indicator lamps. In this application they offer long life, low current drain and small size. But one of their most important applications is in numeric displays, in which arrays of tiny LEDs are arranged to provide illuminated segments that form the numbers. Schematic sym-

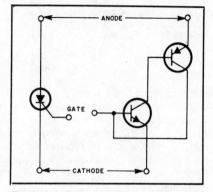

Fig. 6-18 — Schematic symbol for the SCR, along with an equivalent circuit made from two transistors.

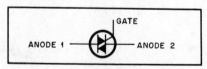

Fig. 6-19 — SCRs are packaged in a wide variety of cases, depending on intended application.

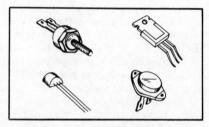

Fig. 6-20 — Schematic symbol for a triac.

Fig. 6-21 — The schematic symbol for an LED is shown at A. At B is a drawing of a typical LED case style.

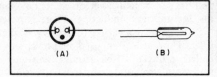

Fig. 6-22 — The schematic symbol for a neon lamp is shown at A. At B is typical NE-2.

bols and typical case styles for the LED are shown in Fig. 6-21.

A typical red LED has a voltage drop of 1.6 V. Yellow and green LEDs have higher voltage drops (2 V for yellow and 4 V for green). The forward-bias current for a typical LED ranges between 10 and 20 mA for maximum brilliance. Bias currents of about 10 mA are recommended for longest device life. As with other diodes, the current through an LED can be varied with series resistors. Varying the current through an LED will affect its intensity; the voltage drop, however, will remain fairly constant.

NEON LAMPS

Neon lamps are used as indicators, or pilot lamps, in amateur equipment. Neon lamps are generally used in the 117-V primary circuit to indicate that the equipment is on. Neon lamps are assigned part numbers with the prefix NE, such as NE-1, NE-2, NE-3 and so on. The NE-2 is the lamp most often found in Amateur Radio equipment.

It takes approximately 67-V dc to light a neon lamp. If the supply voltage is ac, then the peak voltage must be 67 V, so approximately 48-V RMS will fire the lamp. By including about a 150-kΩ resistor in series with one lead, to limit the current through the bulb, you can connect it across the 117-V ac line to use it as a power-on indicator. Another useful feature of the neon lamp is that it will fire in the presence of RF, so these lamps are sometimes used as RF indicators.

Neon lamps are often incorporated into colorful plastic and metal cases to provide a wide variety of panel indicators. Fig. 6-22 depicts a "bare bones" NE-2 lamp and the schematic symbol for a neon lamp.

[Study FCC examination questions with numbers that begin 4AF-4. Review this section as needed.]

CRYSTAL-LATTICE FILTERS

Crystal-lattice filters are used in SSB transmitters and receivers where high-Q, narrow-bandwidth filtering is required. Crystal filters typically are used at intermediate frequencies above 500 kHz in receivers. In SSB transmitters, crystal-lattice filters frequently are used after the balanced modulator to attenuate the unwanted sideband. A quartz crystal acts as an extremely high-Q circuit. The equivalent electrical circuit is depicted in Fig. 6-23A. Fig. 6-23B shows a graph of the reactance vs. frequency for the crystal.

Although single crystals can be used as filtering devices, the normal practice is to wire two or more together in various configurations to provide a desired response curve. Fig. 6-24 depicts a configuration known as the half-lattice filter. In this arrangement, crystals Y1 and Y2 are on different frequencies. The bandwidth and

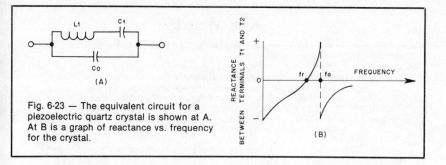

Fig. 6-23 — The equivalent circuit for a piezoelectric quartz crystal is shown at A. At B is a graph of reactance vs. frequency for the crystal.

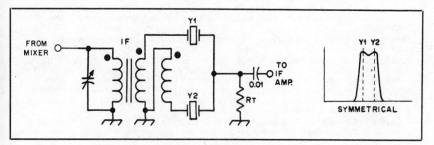

Fig. 6-24 — Schematic diagram and typical response curve of a half-lattice crystal filter.

response shape of the filter depends on the relative frequencies of these two crystals. The overall filter bandwidth is equal to approximately 1 to 1.5 times the frequency separation of the crystals. The closer the crystal frequencies, the narrower the bandwidth of the filter.

A good crystal-lattice filter for DSB voice use would have a bandwidth of approximately 6 kHz at the −6 dB points on the response curve. A good crystal filter for SSB service is significantly narrower; typical bandwidth is 2.1 kHz at the −6 dB points. For CW use, crystal filters typically have 250- to 500-Hz bandwidths.

The home construction of crystal filters can be time consuming and expensive. For this reason, amateurs usually elect to purchase a crystal filter of known performance and base the rest of their design around it. Some amateurs, however, elect to make their own filters. Sometimes, surplus crystals of the right frequency are available. More often than not, however, an amateur will have to locate crystals that are close to the desired frequency, remove them from their cases, and etch or grind them to the right frequency. Generally, it takes a great deal of experimentation to arrive at the right crystal values.

To check your understanding of this last section, study FCC examination questions with numbers that begin 4AF-5. By now, you should have a basic understanding of how all of the circuit components required for your Advanced class license work. If you had trouble with any of the related questions on the FCC Element 4A question pool, review those sections before proceeding to the next chapter.

Key Words

Amplifier transfer function — A graph or equation that relates the input and output of an amplifier under various conditions.

Automatic gain control — An amplifier circuit designed to provide a relatively constant output amplitude over a wide range of input values.

Balanced modulator — A circuit, used to superimpose an information signal on top of an RF carrier, that isolates the input signals from each other and the output, so that only the sum and the difference of the two input signals reaches the output.

Butterworth filter — A filter whose passband frequency response is as flat as possible. The design is based on a Butterworth polynomial to calculate the input/output characteristics.

Chebyshev filter — A filter whose passband and stopband frequency response has an equal-amplitude ripple, and a sharper transition to the stop band than does a Butterworth filter. The design is based on a Chebyshev polynomial to calculate the input/output characteristics.

Constant-k filter — A filter design based on the image-parameter technique. The product of the series and shunt impedances is independent of frequency within the filter passband.

Detector — A circuit used to recover the information signal from a modulated RF envelope.

Doubly balanced mixer (DBM) — A mixer circuit that is balanced for both inputs, so that only the sum and the difference frequencies, but neither of the input frequencies, appear at the output. There will be no output unless both input signals are present.

Elliptical filter — A filter with equal-amplitude passband ripple and points of infinite attenuation in the stop band. The design is based on an elliptical function to calculate the input/output characteristics.

Frequency discriminator — A circuit used to recover the audio from an FM signal. The output amplitude depends on the deviation of the received signal from a center (carrier) frequency.

Half section — A basic L-section building block of image-parameter filters.

Image-parameter technique — A filter-design method that uses image impedance and other fundamental network functions to approximate the desired characteristics.

Linear electronic voltage regulator — A type of voltage-regulator circuit that varies either the current through a fixed dropping resistor or the resistance of the dropping element itself.

L network — A combination of a capacitor and an inductor, one of which is connected in series with the signal lead while the other is shunted to ground.

M-derived filter — A filter, designed using image-parameter techniques, that has a sharper transition from pass band to stop band than a constant-k type. This filter has one infinite attenuation frequency, which can be placed at a desired frequency by proper design.

Mechanical filter — A nonelectrical filter that uses mechanically resonant disks at the design frequency, and a pair of electromechanical transducers to change the electrical signal into a mechanical wave and back again.

Mixer — A circuit that takes two or more input signals, and produces an output that is the sum or difference of those signal frequencies.

Modulator — A circuit designed to superimpose an information signal on an RF carrier wave.

Neutralization — A technique for canceling the positive feedback in an amplifier produced by interelectrode capacitance through the use of a negative-feedback signal supplied by connecting a capacitor from the output to the input circuit.

Parasitics — Undesired oscillations or other responses in an amplifier.

Pi network output-coupling circuits — A combination of two like reactances (coil or capacitor) and one of the opposite type. The single component is connected in series with the signal lead and the two others are shunted to ground, one on either side of the series element.

Product detector — A detector circuit whose output is equal to the product of a beat-frequency oscillator (BFO) and the modulated RF signal applied to it.

Ratio detector — A circuit used to demodulate FM signals. The output is the ratio of voltages from either side of a discriminator-transformer secondary.

Reactance modulator — Used in FM systems, this circuit acts as a variable reactance in an oscillator or amplifier tank circuit.

Slope detection — A method for using an AM receiver to demodulate an FM signal. The signal is tuned to be part way down the slope of the receiver IF filter curve.

Switching regulator — A voltage-regulator circuit in which the output voltage is controlled by turning the pass element on and off at a rate of several kilohertz.

Chapter 7

Practical Circuits

N ow that you have studied some intermediate-level electrical principles, and have learned about the basic properties of some simple components, you are ready to learn how to apply those ideas to practical Amateur Radio circuits. This chapter will lead you through examples and explanations to help you gain that knowledge. In addition to reading the text, you should be prepared to perform all of the calculations shown in the sample problems.

This chapter covers six major types of circuits and how they apply to Amateur Radio. First we will take a look at the principles of voltage-regulator circuits. Next we will study mixers, including modulators and detectors, which are used to put your voice or other information on a radio-wave carrier and recover it again at the receiving end. Then we progress through a study of various amplifier circuits, oscillators, impedance-matching networks and filter circuits. You will be directed to turn to Chapter 10 at appropriate points, and to use the FCC questions as a study aid to review your understanding of the material.

You should keep in mind that there have been entire books written on every topic covered in this chapter. So if you do not understand some of the circuits from our brief discussion, it would be a good idea to consult some other reference books. *The ARRL Handbook* is a good starting point, but even that won't tell you everything about each topic. Our discussion in this chapter should help you understand the circuits well enough to pass your Advanced class exam, however.

ELECTRONIC VOLTAGE-REGULATION PRINCIPLES

Almost every electronic device requires some type of power supply. The power supply must provide the required voltages when the device is operating and drawing a certain current. The output voltage of most power supplies varies inversely with the load current. If the device starts to draw more current, the applied voltage will be pulled down. The operation of most circuits will change as the power-supply voltage changes. Modern solid-state devices are more sensitive to slight voltage changes than many tube circuits are. For this reason, there is a voltage-regulator circuit included in the power supply of almost every electronic device that uses transistors or integrated circuits. The purpose of this circuit is to stabilize the power-supply output voltage and/or current under changing load conditions.

Linear electronic voltage regulators make up one major category of regulator. With these, the regulation is accomplished by varying either the current through a fixed dropping resistance as changes in input voltage or load currents occur, or by varying the resistance of the dropping element. Zener-diode regulator circuits and gaseous-regulator-tube circuits work on the first principle. The latter technique is used in electronic regulators in which the voltage-dropping element is a vacuum tube or a transistor, rather than a resistor. By varying the dc voltage at the grid or the current at the base of these elements, the conductivity of the device may be varied as necessary to hold the output voltage constant. In solid-state regulators, the series-dropping element is called a pass transistor. Power transistors are available which will handle several amperes of current at several hundred volts, but solid-state regulators of this type are usually operated at potentials below 100 V.

The second major regulator category is the *switching regulator*, where the dc source voltage is switched on and off electronically. The average dc voltage available from the regulator is proportional to the duty cycle of the switching wave form, or the ratio of the on time to the total switching-cycle period. Switching frequencies of several kilohertz are normally used, to avoid the need for extensive filtering to smooth the switching frequency from the dc output. We won't go into the operating details of switching regulators in this manual, but you should at least know they exist.

Discrete-Component Regulators

Zener-Diode Shunt Regulators

A Zener diode can be used to stabilize a voltage source, as shown in Fig. 7-1. Note that the cathode side of the diode is connected to the positive supply side. The diode is connected in parallel with, or shunted across, the load. This type of linear voltage regulator is often referred to as a shunt regulator. Zener diodes are available in a wide variety of voltage and power ratings. Voltage ratings range from less than two to a few hundred volts. Power ratings specify the power the diode can dissipate, and run from less than 0.25 W to 50 W. The ability of the Zener diode to stabilize a voltage depends on the diode conducting impedance, which can be as low as 1 ohm or less in a low-voltage, high-power diode, to as high as a thousand ohms in a low-power, high-voltage diode.

Diode Power Dissipation

Zener diodes of a particular voltage rating have varied maximum current capabilities, depending on the diode power ratings. The Ohm's Law relationships you are familiar with can be used to calculate power dissipation, current rating and conducting impedance of a Zener diode.

$$P = I \times E \qquad\qquad\qquad\qquad \text{(Eq. 7-1)}$$

$$I = \frac{P}{E} \qquad\qquad\qquad\qquad \text{(Eq. 7-2)}$$

and

$$Z = \frac{E}{I} \qquad\qquad\qquad\qquad \text{(Eq. 7-3)}$$

where
    P is the diode maximum-safe-power-disipation rating
    E is the Zener voltage
    I is the maximum current that can
        safely flow through the diode
    Z is the diode conducting impedance

How much power is a 50-V Zener diode dissipating if 4 mA flow through it to maintain the regulated output voltage? To solve this problem, use Eq. 7-1:

$$P = 4 \text{ mA} \times 50 \text{ V} = 200 \text{ mW}$$

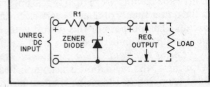

Fig. 7-1 — A voltage-regulator circuit using a Zener diode.

You might also want to change the 4 mA to 0.004 A, then solve. In that case, your answer would be 0.2 W. Of course, you recognize that those answers are equivalent.

If you have a 10-V, 50-W Zener diode operated at its maximum dissipation rating, what is the most current that you can safely allow to flow through the diode? Use Eq. 7-2 to solve for the current:

$$I = 50 \text{ W} / 10 \text{ V} = 5 \text{ A}$$

How much current can a 10-V, 1-W diode safely conduct? The answer to this one is 0.1 A, or 100 mA.

Now calculate the conducting impedance of these diodes. Use Eq. 7-3:

$$10 \text{ V} / 5 \text{ A} = 2 \ \Omega \text{ for the 50-W diode}$$

The 1-W diode has a conducting impedance of 100 Ω. Disregarding small voltage changes that may occur, the conducting impedance of a given diode is inversely proportional to the current flowing through it.

The power-handling capability of most Zener diodes is rated at 25 °C; approximately room temperature. If the diode is operated at a higher ambient temperature, its power-handling capability must be derated. A typical 1-W diode can safely dissipate only ½ W at 100 °C. The breakdown voltage of Zener diodes also varies with temperature. Those rated for operation at 5 or 6 V have the smallest variation with temperature changes, and so are most often used as a voltage reference where temperature stability is a consideration.

Limiting Resistance

The value of R1 in Fig. 7-1 is determined by the load requirements. If R1 is too large the diode will be unable to regulate at large values of I_L, the current through the load, because not enough current will flow through the diode. If R1 is too small, however, the diode dissipation rating may be exceeded when the load current decreases. The optimum value for R1 can be calculated by:

$$R1 = \frac{E_{dc(min)} - E_Z}{1.1 \ I_{L(max)}} \qquad \text{(Eq. 7-4)}$$

Once R1 is known, the maximum diode dissipation for a certain circuit can be determined by:

$$P_D = \left[\frac{E_{dc(max)} - E_Z}{R_1} - I_{L(min)} \right] \times E_Z \qquad \text{(Eq. 7-5)}$$

Eq. 7-4, establishes conditions for the Zener diode to draw 1/10 the maximum load current. Don't use a resistor with a larger value for R1. This will ensure diode regulation under maximum load. For example, assume a 12-V source is to supply a circuit requiring 9 V. The load current varies betweeen 200 and 350 mA. Choose a Zener diode with an E_Z rating of 9.1 V (the nearest available value). Then use Eqs. 7-4 and 7-5 to calculate R1 and the necessary power rating.

$$R1 = \frac{12 \text{ V} - 9.1 \text{ V}}{1.1 \times 0.35 \text{ A}} = \frac{2.9 \text{ V}}{0.385 \text{ A}} = 7.5 \text{ ohms}$$

and

$$P_D = \left[\frac{12 \text{ V} - 9.1 \text{ V}}{7.5} - 0.2 \text{ A} \right] \times 9.1 \text{ V}$$

$$= 0.185 \text{ A} \times 9.1 \text{ V} = 1.7 \text{ W}$$

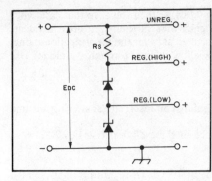

Fig. 7-2 — Zener diodes can be connected in series to obtain multiple output voltages from one regulator circuit.

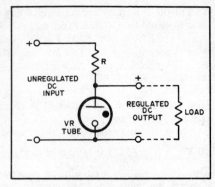

Fig. 7-3 — A voltage-regulator-tube circuit for maintaining a constant output voltage.

It is also a good idea to check the diode dissipation rating with a load current, $I_{L(min)}$, of zero. That way the diode will not be damaged by excess power dissipation even if the load is disconnected. In this case Eq. 7-5 gives a value for P_D of 3.5 W. The nearest available dissipation rating is 5; therefore, a 9.1-V, 5-W Zener diode should be used.

Obtaining Other Voltages

Fig. 7-2 shows how two Zener diodes may be connected in series to obtain regulated voltages that you could not achieve otherwise. This is especially useful when you want to use a 6-V Zener diode for maximum temperature stability, but you need a reference voltage other than 6 V. Two 6-V diodes in series, for example, will provide a reference voltage of 12 V. Another advantage with a circuit of this type is that you have two regulated output voltages. The diodes need not have equal breakdown voltages. You must pay attention to the current-handling capability of each diode, however. The limiting-resistor value can be calculated by Eq. 7-4, if you use the sum of the diode voltages for E_Z and the sum of the load currents for I_L.

Gaseous Regulator Tubes

Occasionally you may need a regulated output voltage at a higher voltage and current rating than can be obtained conveniently with a Zener diode. This may be the case with RF power amplifier circuits. In such applications, gaseous regulator tubes can be used to good advantage. The voltage drop across such tubes is constant over a moderately wide current range. Tubes are available for regulated voltages near 150, 105, 90 and 75 V.

A typical VR tube circuit is shown in Fig. 7-3. You can see that the VR tube is also used as a shunt-type regulator. Let's see how the tube operates. Suppose that the load requires a regulated 150-V source at 25 mA, and that the unregulated dc input voltage is 250 V. Allowing 5 mA (the minimum current for stable operation) through the tube, the total current through the resistor is 5 + 25 = 30 mA. The resistor must drop the supply from 250 V to 150 V, a total of 100 V. Using Ohm's Law, the resistance is approximately 3300 ohms. Now if the input voltage rises to, say, 300 V, R must drop 300 − 150 = 150 V to keep the output voltage constant at 150 V. The current through R, by Ohm's Law, will be 45 mA. The load is still drawing 25 mA, since the voltage across it hasn't changed, so the additional 20 mA must flow through the VR tube. In other words, the VR tube regulates the voltage by drawing more current when the input voltage tends to rise and by drawing less when it tends to decrease.

Series Regulators

The previous section outlines some of the limitations when Zener diodes and VR tubes are used as regulators. Greater current amounts can be accommodated if the Zener diode is used as a voltage reference at low current, permitting the bulk of the load current to flow through a series pass transistor (Q1 of Fig. 7-4). An added benefit in using a pass transistor is that ripple on the output waveform is reduced. This technique is commonly referred to as "electronic filtering."

The pass transistor serves as a simple emitter-follower dc amplifier. It increases the load resistance seen by the Zener diode by a factor of beta (β). In this circuit arrangement, D1 is required to supply only the base current for Q1. The net result is that the load regulation and ripple characteristics are improved by a factor of beta. C1 charges to the peak value of the input voltage ripple, smoothing the dc input to the regulator. The addition of C2 bypasses any remaining hum or ripple around the reference element, reducing the output ripple even more, although many simple supplies such as this do not make use of a capacitor in that part of the circuit. C3 provides some final filtering and helps to stabilize the transistor circuit.

The greater the value of transformer secondary voltage, the higher the power dissipation in Q1. This not only reduces the overall power-supply efficiency, but requires stringent heat sinking at Q1 if the dissipation will be more than a small percentage of the transistor rating.

It is possible to obtain better regulation by adding a few components to monitor the output voltage from your regulated supply. This technique, illustrated in Fig. 7-5, is called feedback regulation. You will notice that an error amplifier detects the difference between the reference voltage and a feedback voltage. This amplified signal is then applied to the base of the pass transistor. The basic circuit operation is the same as for Fig. 7-4. The voltage-divider circuit, connected at the supply output, is adjusted to set a feedback voltage that matches the Zener-diode reference voltage. In this way, you can use any convenient reference diode. It does not have to be the same as the full regulated output voltage.

If the load is connected at the end of a long supply cable, there may be a significant voltage drop when a large current is drawn. This would be the case where the supply is being used to power a solid-state transceiver. By monitoring the voltage at the load rather than at the regulated-supply output, the regulating element can

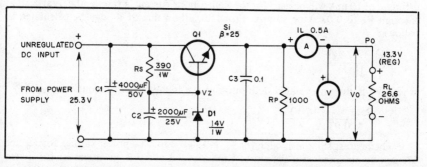

Fig. 7-4 — Illustration of a power-supply regulator including a pass transistor to provide more current than is available from a circuit using only a Zener diode.

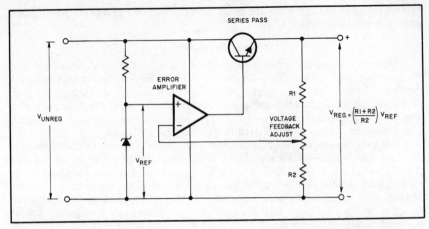

Fig. 7-5 — A sample of the output voltage can be fed back to an error amplifier to obtain better regulation.

respond to the voltage drop in the cable. This technique is called remote-sensed feedback regulation.

Design Example

To calculate the value of R_S in Fig. 7-4, the base current of Q1 must be known. The base current is approximately equal to the Q1 emitter current divided by beta. The transistor beta can be found in the manufacturer's data sheet, or measured with simple test equipment (beta = I_c / I_b). Since the variation in beta for a particular transistor type is a fairly unknown quantity (a 2N3055 is specified to have a beta between 25 and 70), more precise calculations will result if the transistor beta is tested before the calculations are done. A conservative approach is to design for the minimum beta of the transistor used. Calculating I_b for our example,

$$I_b = 0.5 \text{ A} / 25 = 0.02 \text{ A} = 20 \text{ mA}$$

As we pointed out earlier, for D1 to regulate properly it is necessary that the diode draw a fair portion of the current flowing through R_S. The resistor will have 0.02 A of Q1 base current flowing through it, as calculated above. A conservative estimate of 10 mA will be used for the Zener diode current, bringing the total current through R_S to 0.03 A or 30 mA. From this, the value of R_S can be calculated as follows:

$$R_S = \frac{(V' - V_Z)}{I_{RS}} \qquad \text{(Eq. 7-6)}$$

where V' is the unregulated dc input and I_{RS} is the current through R_S.

$$R_S = \frac{(25.3 \text{ V} - 14 \text{ V})}{0.03 \text{ A}} = 376 \ \Omega$$

The nearest standard value for R_S is 390 ohms. The power ratings for R_S and D1 can be obtained with the aid of equations 7-1 and 7-2, given earlier for Zener-diode regulators.

Next we must calculate the power rating for Q1. The power dissipation of Q1 is equal to emitter current times the collector-to-emitter voltage:

$$P_{Q1} = I_E \times V_{CE} \qquad\qquad \text{(Eq. 7-7)}$$

where

$V_{CE} = V' - (V_Z - V_{BE})$, and V_{BE} is approximately 0.6 V for a silicon transistor, and

$P_{Q1} = 0.5 \text{ A} \times 12 \text{ V} = 6 \text{ W}$

It is a good idea to choose a transistor for Q1 that has at least twice the rating calculated. In this example a transistor with a power-dissipation rating of 12 W or more should be used.

Discrete-Component-Regulator Limitations

Shunt-regulator circuits are inherently inefficient because the regulating element draws maximum current when the load is drawing none. This type of circuit is most useful if the load current will be fairly constant, or if you want to maintain a nearly constant load on the unregulated supply. The current drawn by the shunt element represents wasted power, however. A series-regulator circuit will make more efficient use of the unregulated supply, because it draws a minimum current when the load current is zero.

The primary limitation of using a pass transistor is that it can be destroyed almost immediately if a severe overload occurs at R_L. A fuse cannot blow fast enough to protect Q1 in case of an accidental short circuit at the output, so a current-limiting circuit is required. An example of a suitable circuit is shown in Fig. 7-6.

All of the load current is routed through R2. There will be a voltage difference across R2 that depends on the exact load current at a given instant. When the load current exceeds a predetermined safe value, the voltage drop across R2 will forward bias Q2, causing it to conduct. If you select a silicon transistor for Q2 and a silicon diode for D2, the combined voltage drops through them (roughly 0.6 V each) will be 1.2 V. Therefore, the voltage drop across R2 must exceed 1.2 V before Q2 can turn on. Choose a value for R2 that provides a drop of 1.2 V when the maximum safe load current is drawn. In this example, there will be a 1.2-V drop across R2 when I_L reaches 0.43 A.

When Q2 turns on, some of the current through R_S flows through Q2, thereby depriving Q1 of some of its base current. Depending on the amount of Q1 base cur-

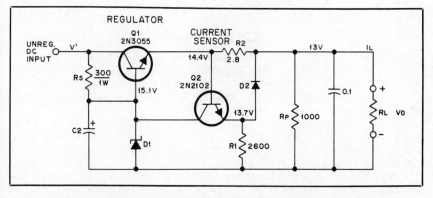

Fig. 7-6 — Overload protection for a regulated supply can be effected by addition of a current-sense transistor to turn off the regulator in case of a short circuit on the output.

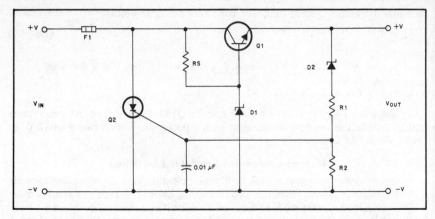

Fig. 7-7 — An SCR provides a "crowbar" feature to short the unregulated supply and blow the fuse if the pass transistor fails.

rent at a precise moment, this action cuts off Q1 conduction to some degree, thus limiting the current flow through it.

If the collector-emitter junction of the pass transistor becomes shorted, the full unregulated rectifier voltage can be applied to the load. This can lead to a disaster, especially if the load is an expensive piece of electronic equipment! So you see, the load can draw too much current and damage the supply, or the supply can provide too high a voltage, and damage the load. The current-sense transistor protects the supply, and a "crowbar circuit" is often used to protect the load.

Fig. 7-7 shows a simple crowbar circuit. A silicon-controlled rectifier (SCR) is selected that will turn on when a set voltage is applied to its gate terminal. Zener diode D1, and resistors R1 and R2, are used to sense a voltage for the SCR. When the set-point voltage is exceeded, the SCR turns on, creating a short circuit across the power-supply output terminals. This sustained short circuit will blow a fuse, turning the supply off. The key is that the SCR "fires" quickly, shorting the output voltage before it can damage the load equipment.

Just as it is possible to add several Zener diodes in series to increase the regulated output voltage, you can connect pass transistors in parallel to increase the current-handling capability of the regulator. You can include current limiting, overvoltage protection, remote-sensed feedback regulation and other features in a single supply.

IC Regulators

The modern trend in regulators is toward the use of integrated-circuit devices known as three-terminal regulators. Inside these tiny packages is a voltage reference, a high-gain error amplifier, current-sense transistors and resistors, and a series pass element. Some of the more sophisticated units have thermal shutdown, overvoltage protection and foldback current limiting.

Three-terminal regulators have a connection for unregulated dc input, one for regulated dc output and one for ground. They are available in a wide range of voltage and current ratings. It is easy to see why regulators of this sort are so popular when you consider the low price and the number of individual components they can replace. The regulators are available in several different package styles. The package and mounting methods you choose will depend on the amount of current required from your supply. The larger metal TO-3 package, mounted on a heat sink, for example, will handle quite a bit more current than a plastic DIP IC.

Three-terminal regulators are available as positive or negative types. In most cases, a positive regulator is used to regulate a positive voltage and a negative regulator for a negative voltage (with respect to ground). However, depending on the system ground requirements, each regulator type may be used to regulate the "opposite" voltage.

Parts A and B of Fig. 7-8 illustrate the conventional method of connecting an IC regulator. Several regulators can be used with a common-input supply to deliver a variety of voltages with a common ground. Negative regulators may be used in the same manner, to provide several negative voltages, or with positive regulators to provide supplies with both positive and negative polarities.

Parts C and D of Fig. 7-8 show how a regulator can be connected to provide opposite-polarity voltages, as long as no other supplies operate from the unregulated input source. In these configurations the input supply is floated; neither side of the input is tied to the system ground.

When choosing a three-terminal regulator for a given application, the important specifications to look for are maximum output current, maximum output voltage, minimum and maximum input voltage, line regulation, load regulation and power dissipation.

In use, most of these regulators require an adequate heat sink, since they may be called on to dissipate a fair amount of power. Also, since the chip contains a high-gain error amplifier, bypass capacitors on the input and output leads are essential for stable operation. Most manufacturers recommend bypassing the input and output directly at the IC leads. Tantalum capacitors are usually recommended because of their excellent bypass capabilities up into the VHF range.

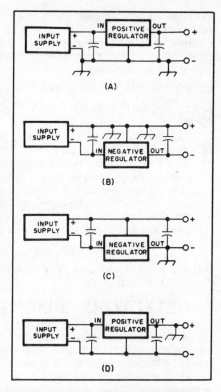

Fig. 7-8 — Parts A and B illustrate the conventional manner of connecting three-terminal regulators. Parts C and D show how one regulator polarity can be used to provide an output voltage of the opposite polarity.

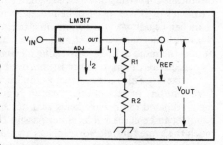

Fig. 7-9 — By varying the ratio of R2 to R1 in this simple regulator circuit, a wide range of output voltages is possible.

Adjustable-Voltage IC Regulators

Adjustable-voltage regulators are similar to the fixed-voltage devices just described. The main difference is that a fourth lead is provided to connect the error-amplifier sense voltage to a voltage divider. See Fig. 7-9. This allows you to select the portion of the output voltage that is fed back to the amplifier, thus providing an output voltage to suit your needs, at least within certain limits. The ratio of reference voltage to total output voltage will be the same as the ratio R2 / (R2 + R1). For example, the regulator shown in Fig. 7-9 (the LM-317 is a 1.2- to 37-V adjustable regulator) requires a reference voltage of 1.2 V. To use this regulator to build a 12-V supply, the voltage ratio is 1.2 V / 12 V = 0.1. If we select R2 = 1 kΩ, then

R2 + R1 = 1000 Ω / 0.1 = 10,000 Ω,

so R1 = 10,000 Ω − 1000 Ω = 9000 Ω. Select a 9.1-kΩ standard-value component for R1.

Regulators are readily available to provide an output voltage of from 5 to 24 V at up to 5 A. The same precautions should be taken with these types of regulators as with the fixed-voltage units. Proper heat sinking and lead bypassing is essential for proper circuit operation.

It is a rather simple task to design a regulated power supply around an IC regulator. Multiple voltages can be obtained by simply adding other ICs in parallel. The main consideration is to provide an input voltage within the range specified for the device you select, and to be sure your load will not draw more current than the IC can safely supply.

[Now turn to Chapter 10 and study FCC examination questions with numbers that begin 4AG-1 and 4AG-12. Review this section as needed.]

DETECTORS, MIXERS AND MODULATORS

A *detector* circuit is used to "reclaim" the information that has been superimposed on a radio wave at a transmitter. That information may be the operator's voice, or it may be Morse code or RTTY signals. It could even be the information that will allow you to reproduce a slow-scan TV or facsimile picture that has been transmitted across the miles. There are a variety of detector methods described in this section, but they are all associated with receivers. If the circuit introduces a signal loss during processing, it is called a passive detector. If some gain is provided (or at least no signal loss) then the circuit is called an active detector.

Mixer circuits are used to change the frequency of the desired signal. In a receiver, this means converting all received signals to a single frequency (called the intermediate frequency or IF) so they can be processed more efficiently. In this way the amplifier chain can be adjusted for maximum efficiency and the best signal-handling characteristics without the need to retune many circuit elements every time you change bands or turn the radio dial. By converting all received signals to a single IF for processing, the selectivity of the receiver is improved greatly.

Mixers are also used to change the frequency of a signal as it progresses through a transmitter. In fact, a mixer circuit is even used in most transmitters to produce the modulated radio wave being sent out across the airwaves. Like detectors, mixer circuits can be classified as either passive or active. The principles of operation are much the same for detectors, mixers and *modulators*, so when you become familiar with the operation of one type, you will find the others easier to understand.

Mixers

When two sine waves are combined in a nonlinear circuit, the result is a complex waveform that includes the two original signals, a sine wave that is the sum of the two frequencies and a sine wave that is the difference between the two. Also included are combinations of the harmonics from the input signals, although these are usually weak enough to be ignored. One of the product signals can be selected at the output by using a filter. Of course, the better the filter, the less of the unwanted products or the two input signals that will appear in the final output signal. By using

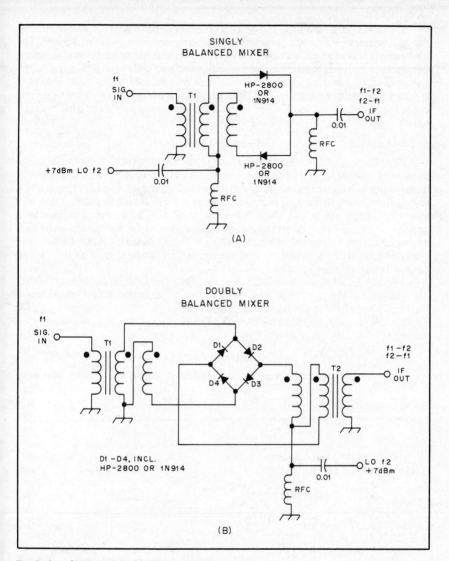

Fig. 7-10 — Singly and doubly balanced diode mixers.

balanced-mixer techniques, the mixer circuit provides isolation of the various ports, so the signals at the RF, LO and IF ports will not appear at any other. This prevents the two input signals from reaching the output. In that case the filter needs to remove only the unwanted mixer product.

In a typical amateur receiving application, you want to mix the RF energy of a desired signal with the output from an oscillator in the receiver so you can produce a specific output frequency (IF) for further processing in the radio. The local oscillator (LO) frequency determines the frequency that the input signal is converted to. By using this frequency-conversion technique, the IF stages can be designed to operate over a relatively narrow frequency range, and filters can be designed to provide a high degree of selectivity in the IF stages.

The mixer stages in a high-performance receiver must be given careful considera-tion. These stages will have a great impact on the dynamic range of the receiver. The RF signal should be amplified only enough to overcome mixer noise. Otherwise, strong signals will cause desensitization, cross-modulation and IMD products in the mixer. The level of spurious mixer products may also be increased to the point that they appear in the output. One result of these effects is that the receiver may be useless in the presence of extremely strong signals. A mixer should be able to handle strong signals without being affected adversely.

Passive Mixers

One simple mixer that has good strong-signal characteristics is the diode mixer shown in Fig. 7-10A. This circuit makes use of a trifilar-wound broadband toroidal transformer to balance the mixer, effectively canceling the RF and LO input signals, so only the sum and difference frequencies appear at the IF output terminal. Sometimes this circuit is referred to as a singly balanced mixer. Fig. 7-10B shows an improved circuit that includes two balance transformers and four diodes arranged in a circular, or ring, pattern. This circuit is usually called a *doubly balanced mixer (DBM)* or a diode-ring mixer. Note especially that the diodes are not connected as they would be in a bridge rectifier circuit.

The doubly balanced mixer is the most common. Commercial modules offer elec-trical balance at the ports that would not be easy to achieve with homemade trans-formers. They also use diodes whose characteristics have been carefully matched. Typical loss through a DBM is 6 to 9 dB. The port-to-port isolation is usually on the order of 40 dB.

Active Mixers

While passive mixers have good strong-signal-handling ability, they also have

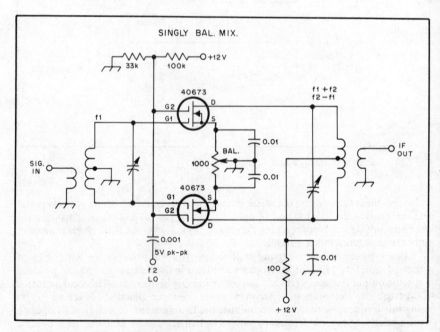

Fig. 7-11 — An active singly balanced FET mixer.

some drawbacks. They require a relatively strong LO signal, and they generate a fair amount of noise. Active mixers can be used to advantage if you require less conversion loss, weaker LO signals and less noise. Just be aware that the strong-signal-handling capabilities are not generally as good.

A JFET or dual-gate MOSFET can be used as a mixer, and will provide some gain as well as mixing the signals for you. Bipolar transistors could be used, but seldom are. Fig. 7-11 shows an active mixer circuit. Many variations are possible, and this diagram just shows one arrangement.

Integrated-circuit mixers are available in both singly and doubly balanced types. These devices provide at least several decibels of conversion gain, low noise and good port-to-port isolation.

Detectors

The simplest type of detector, used in the very first radio receivers, is the diode detector. A complete, simple receiver is shown in Fig. 7-12. This circuit would only work for strong AM signals, so it is not used very much today, except for experimentation. It does serve as a good starting point to understand detector operation, however. The waveforms shown on the diagram illustrate the changes made to the signal as it progresses through the circuit. L1 couples the received RF signal to the tuning circuit of L2-C1. The diode rectifies the RF waveform, passing only the positive half cycles. C2 charges to the peak voltage of each pulse, producing a smoothed dc waveform, which then goes through the voltage divider R1 and R2. R2 selects a portion of the signal voltage to be applied to the headphones, providing a volume-control feature. C4 is a coupling capacitor that serves to remove the dc offset voltage, leaving an ac audio signal.

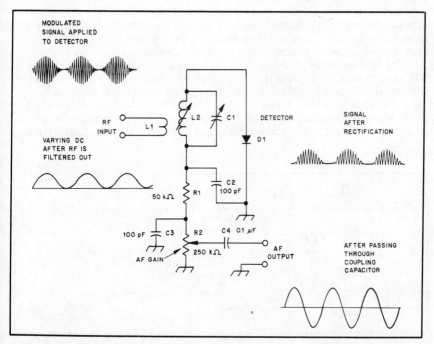

Fig. 7-12 — A simple receiver circuit using a single diode detector. L2/C1 is tuned to the desired receive frequency.

One drawback of the diode detector is that there is some signal loss in the circuit. An FET can be used as a detector, and the transistor will provide some amplification. The disadvantage of an active detector is that it may be overloaded by strong signals more easily than the passive diode detector.

Product Detectors

A *product detector* is similar to a balanced mixer. It is a detector whose output is equal to the product of a beat-frequency oscillator (BFO) and the RF signal. The BFO frequency is chosen to provide detector output at audio frequencies. A doubly balanced diode-ring product detector looks very much like the doubly balanced diode-ring mixer shown in Fig. 7-10B. FETs can be used for product detectors, and those circuits also look much the same as the mixer circuits. Special IC packages are also available for use as product detectors, and they will provide the advantages of diode detectors in addition to several decibels of gain. Product detectors are used for SSB, CW and RTTY reception.

Detecting FM Signals

There are three common ways to recover the audio information from a frequency-modulated signal. A *frequency-discriminator* circuit uses a transformer with a center-tapped secondary. The primary signal is introduced to the secondary-side center tap through a capacitor. With an unmodulated input signal, the secondary voltages either side of the center tap will cancel. But when the signal frequency changes, there is a phase shift in the two output voltages. These two voltages are rectified by a pair of diodes, and the resulting signal varies at an audio rate. Fig. 7-13 shows the schematic diagram of a simple frequency discriminator. Crystal descriminators use a quartz-crystal resonator instead of the LC tuned circuit in the frequency discriminator, which is often difficult to adjust properly.

A *ratio detector* can also be used to receive FM signals. The basic operation depends on the rectified output from a transformer similar to the one used in a frequency discriminator being split into two parts by a divider circuit. Then it is the ratio between these two voltages that is used to produce the audio signal.

The last method of receiving an FM signal can be used if you have an AM or SSB receiver capable of tuning the desired frequency range. If you tune the receiver slightly off the center frequency, so the varying signal frequency moves up and down the slope of the selectivity curve, an AM signal will be produced. (This technique is also known as *slope detection*.) This AM signal then proceeds through the receiver in the normal fashion, and you can hear the audio in the speaker. Careful tuning

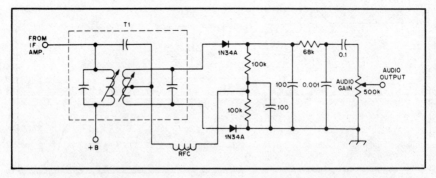

Fig. 7-13 — A typical frequency-discriminator circuit used for FM detection. T1 is a Miller 12-C45 discriminator transformer.

will make the FM signal perfectly understandable, although you might have to keep one hand on the tuning knob, and there may be some noise.

Modulator Circuits

Modulation is really a mixing process; any mixer or converter circuit could be used for generating a modulated signal. Instead of introducing two radio frequencies into a mixer circuit, we simply introduce one radio frequency (the carrier frequency) and the voice-band audio frequencies.

Mixer circuits used in receivers are designed to handle a small signal and a large local-oscillator voltage. This means that the percentage of modulation of the IF signal is low. In a transmitter we want to get as close as possible to 100% modulation, and we also want more power output. For these reasons modulator circuits differ in detail from receiving mixers, although they are much the same in principle.

Amplitude Modulation

Although double-sideband, full-carrier, amplitude modulation (DSB AM) is seldom used on the amateur bands anymore, we will discuss a simple system just to help you understand how a single-sideband, suppressed-carrier (SSB) signal is generated. When RF and AF signals are combined in a DSB AM transmitter, four output signals are generated. First, there is the original audio signal, which is easily rejected by the tuned RF circuits in the stages following the modulator. Then there is the RF carrier, also unchanged from its original form. Of primary importance are the other two signals: one being the sum of the carrier frequency and the audio, the other being the difference between the original signals. These two new signals are called sideband signals. The amplitude of these signals at any given instant depends on what the amplitude of the original audio signal is at that instant. The greater the audio-signal strength, the greater the amplitude of the sideband signals.

The sum component is called the upper sideband. As the audio-signal frequency increases, so does the frequency of the upper-sideband signal. The difference component is called the lower sideband. This sideband is inverted, which means that as the audio-signal frequency increases, the sideband frequency decreases.

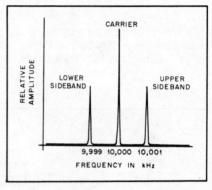

Fig. 7-14 — A 10-MHz carrier, modulated by a 1-kHz sine wave, produces an output as shown.

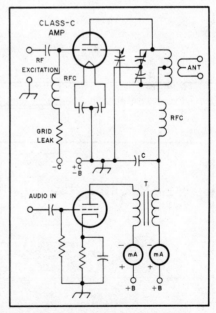

Fig. 7-15 — Plate modulation of a class-C RF amplifier.

The result of all this is that the RF envelope, as viewed on an oscilloscope, has the general shape of the modulating waveform. The envelope varies in amplitude because it is the vector sum of the carrier and the sidebands. Fig. 7-14 illustrates this principle. Note that the carrier amplitude is not changed in amplitude modulation. This is a common misunderstanding, one that often leads to confusion. It is the RF envelope that varies in amplitude.

You can produce amplitude modulation by applying the AF signal to the plate or collector voltage of an RF amplifier stage, as shown in Fig. 7-15. You can also modulate the control element of the amplifier (grid, base or gate). A wide variety of modulator circuits have been used over the years.

SSB: The Filter Method

One way to generate an SSB signal is to remove the carrier and one of the sidebands from an ordinary DSB AM signal. The block diagram shown in Fig. 7-16 shows how it is done. The RF oscillator generates a carrier wave that is injected into a

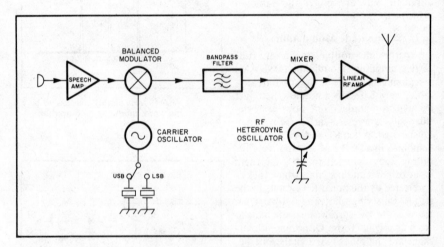

Fig. 7-16 — A block diagram showing the filter method of generating an SSB signal.

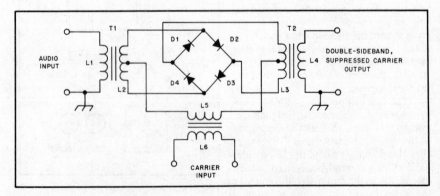

Fig. 7-17 — One example of a diode-ring doubly balanced modulator.

balanced modulator. With this system, the oscillator frequency is changed, depending on which sideband you want to use. Another system maintains the same oscillator frequency, but switches filters to remove the opposite sideband. The audio information is amplified by the speech amplifier and is then applied to the modulator. A balanced modulator takes these two inputs and supplies, as its output, both sidebands without the carrier. This meets the first requirement for the generation of an SSB signal — removal of the carrier.

Let's see how the balanced modulator accomplished this. There are many different types of balanced modulators and it would be impossible to show them all here. One of the more popular types is illustrated in Fig. 7-17. This particular circuit is called a diode-ring balanced modulator. If you get the feeling that you have seen all this before, don't worry. The diode-ring balanced modulator is very similar to the doubly balanced diode-ring mixer. Actually, you could redraw Fig. 7-17 to look almost identical to Fig. 7-10B.

Audio information is coupled into the circuit through transformer T1. The carrier is injected through coils L5 and L6 and the double-sideband, suppressed-carrier output is taken through L3 and L4. To better understand the circuit operation, let's first analyze the circuit with only the carrier applied. See Fig. 7-18A. The polarity of voltage shown across L5 will cause current to flow in the direction indicated by the arrows. D1 and D3 will conduct. The current that flows through each half of L3 is equal and opposite, causing a canceling effect. Output at L4 will be zero. During the next half cycle, the polarity of voltage across L5 will reverse, D2 and D4 will conduct, and again the output at L4 will be zero.

Now for a moment, let's remove the carrier signal and connect an audio source to the audio-input terminals. Fig. 7-18B illustrates this condition. During one half of the audio cycle, D2 and D3 will conduct and the output will be zero. On the other half cycle of the audio signal, D1 and D4 will conduct and again the output at L4 will be zero. Notice that in this case, no signal flows through L3 at all. We can see that if either the carrier or the audio is applied without the other, there will be no output.

Both signals must be applied if the circuit is to work as intended. In practice, the carrier level is made much larger than the audio input. Diode conduction is, therefore, determined by the carrier. There are four conditions that can exist as far as the voltage polarity of the carrier and audio signals are concerned. Both waves can be positive, both can be negative or they can be opposites. We must keep in mind that the carrier is going through many cycles while the audio sine wave goes through only one.

Let's look again at the drawing at Fig. 7-18C. With the polarities indicated, the major current flow will be through D3 since the audio and carrier voltages are aiding (adding together) in this path. Since the balance through L3 has been upset (the bottom half of L3 has more current flowing through it than does the top half), there will be an output present at L4. Next, the carrier polarity will reverse while the audio polarity remains the same. This is the order in which the polarities would change, since the carrier is reversing polarity at a much faster rate than the audio signal. Later, the carrier polarity will be back to what it was at C; however, now the audio will be on the negative portion of its sine wave and so its polarity will be reversed. Maximum current flow under these conditions is through D1 and again an output signal appears at L4. The fourth condition that will exist is shown when the audio-signal polarity is the same as in C, but the carrier is reversed. This time, D4 will be the main path for current flow. As you have probably guessed, there is output at L4 under these conditions.

Fig. 7-19 shows a composite drawing of the audio and carrier waveforms. The shaded areas represent the double-sideband, suppressed-carrier output from the balanced modulator.

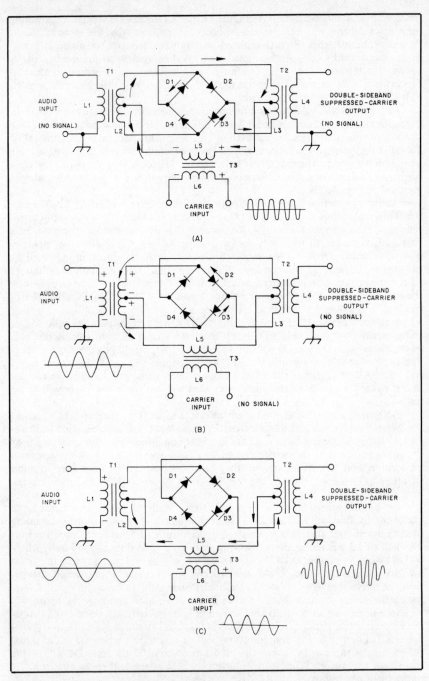

Fig. 7-18 — Arrows indicate the current direction in a diode-ring mixer circuit. Part A shows the condition when only a carrier signal is applied. B is the current with audio applied but no carrier, and C shows the current for one polarity relationship of carrier- and audio-input signals. See the text for more information about the other three possible polarity relationships.

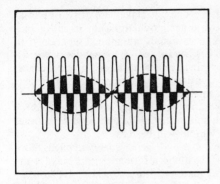

Fig. 7-19 — Superimposed audio and RF waveforms. The shaded area represents the double-sideband, suppressed-carrier output from a balanced modulator.

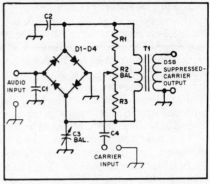

Fig. 7-20 — Here is another common form of balanced modulator. C3 and R2 are adjusted for maximum carrier suppression.

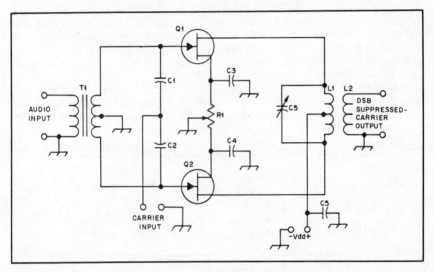

Fig. 7-21 — This circuit is an active balanced modulator using two FET devices. R1 is adjusted for maximum carrier suppression, and C5 is adjusted for maximum double-sideband suppressed-carrier output.

Another type of diode-ring balanced modulator is shown in Fig. 7-20. Two balance controls are provided so that the circuit can be adjusted for optimum carrier suppression (50 dB is a practical amount). This circuit operates basically the same way the modulator shown in Figs. 7-17 and 7-18 works. With either the audio or carrier applied separately, the circuit is balanced and there will be no output. With both the audio and carrier applied, the balance is upset and output will be present at T1. We won't go into the detail that we did in the previous circuit. You should be able to determine which diodes are conducting for the different voltage polarities presented by the audio and carrier signals.

The two circuits we have examined can be classified as passive balanced modulators. That is, they do not provide gain (amplify) but acutally cause a small amount of signal loss (insertion loss). Balanced modulators can be built using active devices.

Practical Circuits   7-19

One such modulator is shown in Fig. 7-21. This circuit makes use of two FETs as the active devices. As was true with the two other modulators, the circuit is in a balanced condition if either the audio or carrier signals are applied separately. In this circuit the tuned output network (C5/L1) is adjusted for resonance at the RF carrier frequency. At audio frequencies, this circuit represents a very low impedance, allowing no audio to appear at the output. Consider the carrier input for a moment. Injection voltage is supplied to each gate in a parallel fashion. The input to each gate is of equal amplitude and of the same phase, and the output circuit is connected for push-pull operation. Currents through each half of the tank are equal and opposite. The signals will effectively cancel and the output will be zero.

Let's analyze the circuit with both the audio and carrier energy applied. Since we have a push-pull input arrangement for the audio information, the bias for the FETs varies at an audio rate. The audio signal applied to one gate is 180° out of phase with the other. While one of the devices is forward biased, the other is reverse biased. The input to each device is the audio signal plus the carrier signal. Sum and difference frequencies are developed at the output to produce a double-sideband, suppressed-carrier signal.

Removing the Unwanted Sideband

We now have a signal that contains both the upper and lower sidebands of what was an amplitude-modulated signal with carrier. The next step in generating our SSB signal is to remove one of the sidebands. Looking back at Fig. 7-16, we see that the next stage after the balanced modulator is the filter. This circuit does just as its name suggests — it filters out one of the sidebands.

An example of a simple crystal filter is shown in Fig. 7-22. The two crystals would be separated by approximately 2 kHz to provide a bandwidth suitable for passing one sideband and not the other. The curve to the right of the filter shows the shape

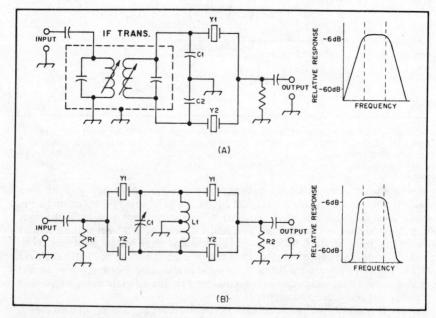

Fig. 7-22 — A half-lattice band-pass filter is shown at A. B shows two half-lattice filters in cascade. The filter response is shown to the right of each filter.

of the filter response. More-elaborate filters using four and six crystals will give steeper slopes on the response curve, without affecting the bandwidth near the top of the curve. As shown at B, a filter with more crystals has a narrower response at the − 60 dB point. The filter at B has better "skirt selectivity" than the one at at A. Two half-lattice filters of the type shown at A are connected back to back to form the filter at B. Crystal-lattice filters of this type are available commercially for frequencies up to 40 MHz or so.

Amplification for the Modulator

The last stage shown in the block diagram in Fig. 7-16 is the linear amplifier. Since the modulation process occurs at low power levels in a conventional transmitter, it is necessary that any amplifiers following the balanced modulator be linear. In the low-power stages of a transmitter, high voltage gain and maximum linearity are quite a bit more important than efficiency.

Generating an FM Signal

Any type of modulation that changes the phase angle of the transmitted sine wave is called angle modulation. Frequency modulation and phase modulation are the two common types of angle modulation. Most methods of producing FM will fall into two general catagories. They are direct FM and indirect FM. As you might expect, each has its advantages and disadvantages. Let's look at the direct-FM method first.

Direct FM

A *reactance modulator* is a simple and satisfactory device for producing FM in an amateur transmitter. This is a vacuum tube or transistor connected to the RF tank circuit of an oscillator so that it acts as a variable inductance or capacitance. The only way to produce a true emission F3E signal is with a reactance modulator on the transmitter oscillator.

Fig. 7-23 is a representative circuit. Gate 1 of the modulator MOSFET is connected across the oscillator tank circuit (C1 and L1) through resistor R1 and blocking capacitor C2. C3 represents the input capacitance of the modulator transistor.

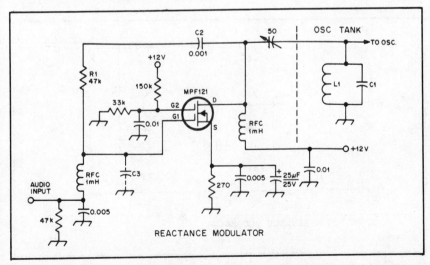

Fig. 7-23 — A reactance modulator using a high-transconductance MOSFET.

R1 is made large compared to the reactance of C3, so the RF current through R1 and C3 will be practically in phase with the RF voltage appearing at the terminals of the tank circuit. The voltage across C3 will lag the current by 90°, however. The RF current in the drain circuit of the modulator will be in phase with the gate voltage, and consequently is 90° behind the current through C3, or lagging the RF-tank-circuit voltage by 90°. This lagging current is drawn through the oscillator tank, giving the same effect as though an inductor were connected across the tank. The frequency increases in proportion to the amplitude of the lagging modulator current. The audio voltage, introduced through a radio-frequency choke, varies the transconductance of the transistor and thereby varies the RF drain current.

The modulator sensitivity (frequency change per unit change in modulating voltage) depends on the transconductance of the modulator transistor. Sensitivity increases when R1 is made smaller in comparison with the reactance of C3. It also increases with an increased L/C ratio in the oscillator tank circuit. For highest carrier stability, however, it is desirable to use the largest tank capacitance that will permit you to obtain the required deviation, while keeping within the limits of linear operation.

A change in any of the modulator-transistor voltages will cause a change in RF drain current, and consequently, a frequency change. Therefore, you should use a regulated power supply for both modulator and oscillator.

A reactance modulator can be connected to a crystal oscillator, as shown in Fig. 7-24. The resulting signal will be more phase modulated than frequency modulated, however, since varying the frequency of a crystal oscillator will only produce a small amount of frequency deviation. Notice that this particular circuit uses a varactor diode to change the circuit capacitance by means of a bias voltage and the modulating signal. So, you should recognize that a reactance modulator acts as either a variable inductor or a variable capacitor in the tank-circuit oscillator.

Fig. 7-25 is a block diagram of a system that uses a reactance modulator to shift an oscillator frequency and generate an FM signal directly. Successive multiplier stages provide output on the desired frequency, which is then amplified by a power amplifier stage.

With any reactance modulator, the modulated oscillator is usually operated on

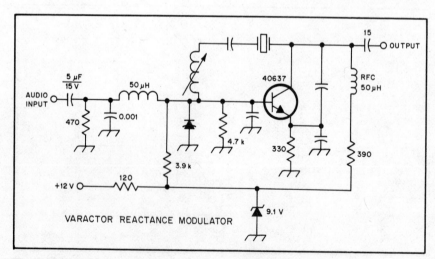

Fig. 7-24 — A reactance modulator using a varactor diode.

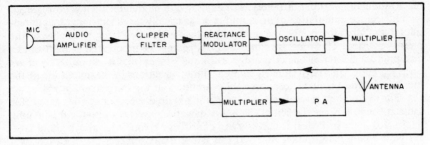

Fig. 7-25 — Block diagram of a direct-FM transmitter.

a relatively low frequency, so that a high order of carrier stability can be secured. Frequency multipliers are used to provide the final desired output frequency. It is important to note that when the frequency is multiplied, so is the frequency deviation. So the amount of deviation produced by the modulator must be adjusted carfully to give the proper deviation at the final output frequency.

Indirect FM

The same type of reactance-modulator circuit that is used to vary the oscillator-tank tuning for an FM system can be used to vary the amplifier-tank tuning, and thus vary the phase of the tank current, to produce phase modulation. Hence, the modulator curcuit of Fig. 7-24 or Fig. 7-25 can be used for PM (emission G3E) if the reactance transistor or tube works on an amplifier tank instead of directly on a self-controlled oscillator.

The phase shift that occurs when a circuit is detuned from resonance depends on the amount of detuning and the circuit Q. The higher the Q, the smaller the amount of detuning needed to secure a given amount of phase shift. If the Q is at least 10, the relationship between phase shift and detuning (in kilohertz either side of the resonant frequency) will be substantially linear over a phase-shift range of about 25 °. From the standpoint of modulator sensitivity, the Q of the tuned circuit on which the modulator operates should be as high as possible. On the other hand, the effective Q of the circuit will not be very high if the amplifier is delivering power to a load, since the load resistance reduces the Q. There must, therefore, be a compromise between modulator sensitivity and RF power output from the modulated amplifier. An optimum Q figure appears to be about 20; this allows reasonable loading of the modulated amplifier, and the necessary tuning variation can be secured from a reac-

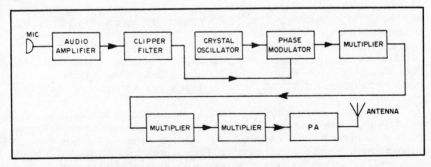

Fig. 7-26 — Block diagram of an indirect-FM transmitter.

tance modulator without difficulty. It is advisable to modulate at a low power level.

Reactance modulation of an amplifier stage usually results in simultaneous amplitude modulation because the modulated stage is detuned from resonance as the phase is shifted. This must be eliminated by feeding the modulated signal through an amplitude limiter or one or more "saturating" stages; that is, amplifiers that are operated class C and driven hard enough so that variations in the amplitude of the input excitation produce no appreciable variations in the output amplitude.

For this type of reactance modulator, the required speech-amplifier gain is the same as for FM. Since the actual frequency deviation increases with the modulating audio frequency in PM, it is necessary to cut off the frequencies above about 3000 Hz before modulation takes place. If this is not done, unnecessary sidebands will be generated at frequencies considerably removed from the carrier frequency.

The indirect method of generating FM shown in Fig. 7-26 is currently popular. Shaped audio is applied to a phase modulator to generate PM. Since the amount of deviation produced is very small, a large number of multiplier stages is necessary to achieve wideband deviation at the operating frequency.

[Study FCC examination questions with numbers that begin 4AG-6 and 4AG-8. Review this section as needed.]

AMPLIFIERS

The energy picked up by a receiving antenna is extremely small. Therefore, the strength of the signals must be increased by a large factor in a receiver — often a million or more times. You might think that this amplification process could be carried on indefinitely, so that even the weakest radio signals could be brought up to usable strength. Unfortunately, there are limitations on how much you can amplify a signal.

Noise across the radio spectrum limits the amount of useful amplification. Unlimited amplification of a particular frequency cannot reduce the noise mask covering a desired signal, since both are RF emissions. Where does this noise come from? Electrical currents generated in nature (static) and in electrical circuits and devices all produce noise. Occurring at all frequencies, this noise is inescapable.

Amateurs don't give up in their battle against noise, though, because there are means to make it less bothersome. One of the big objectives in designing good receivers is to improve the signal-to-noise ratio. This means amplifying the signal as much as possible while amplifying the noise as little as possible.

Amplifier Operating Class

The basic operation of an amplifier is selected by choosing a bias or dc input when there is no signal applied to the amplifier. In the most common amplifier configuration, tubes and PNP transistors are operated with a slightly negative bias voltage applied to the grid or base, and NPN transistors with a slightly positive bias voltage applied to the base. The actual bias voltage depends on the amplifying device used and the type of amplification desired.

The amplifier operating characteristics define several classes of operation. The three basic operating classes are class A, class B and class C. Another class of operation is sometimes added between the A and B categories, and this is called class-AB operation.

To understand what we mean by operating class, let's consider something called an *amplifier transfer function*. This is simply a way of expressing the amplifier output in terms of the input. All amplifiers have their own transfer function, but they all have certain characteristics in common. There is usually a point at which a bigger input signal will not produce a bigger output signal. This is called the saturation region.

There is also a point at which the output will not decrease if a weaker (or no) input signal is supplied. This is called the cutoff region. Between saturation and cutoff is an area where the input and output signals vary in a linear fashion. Of course, this is called the linear region. Fig. 7-27 illustrates a typical transfer function, showing the important regions on the curve.

Two factors about the input are extremely important for establishing the type of operation required of a specific amplifier. The first, called the bias, is the average signal input level, or the dc input level. The second is the size of the input signal. These levels must be set carefully to produce the desired type of operation.

With a class-A amplifier, the bias level and input-signal amplitude are set so that all of the input signals appear between the saturation and cutoff regions. This means the amplifier is operating in the linear region of its transfer function, and the output is a linear (but larger) reproduction of the input signal. There is an output signal for the full 360° of input signal. That portion of the input cycle for which there is an output is called the conduction angle. For a class-A amplifier, the conduction angle is 360°. Fig. 7-28A is a graph of the output current from a class-A amplifier. The efficiency of a class-A amplifier is low, because there is always a significant amount of current drawn from the power supply — even with no input signal. This no-signal current is called the quiescent current of the amplifier. The maximum theoretical efficiency of a class-A amplifier is 50%, but in practice it is more like 25 to 30%.

For a class-AB amplifier, the drive level and dc bias are adjusted so output current flows for more than half the input cycle, but not for the entire cycle. The conduction angle is between 180° and 360°, and the operating efficiency is often more than 50%. Fig. 7-28B shows the output signal for a class-AB amplifier.

Class-B operation sets the bias right at the cutoff level. In this case, output current flows only during half the input sine wave. This represents a conduction angle of 180°. See Fig. 7-28C The output is not as linear as with a class-A amplifier, but is still acceptable for many applications. The advantage is increased efficiency; up to 65% efficiency is possible with a class-B amplifier.

Class-C amplification requires that the bias be well below cutoff and that the signal be large enough to bring some part of the top half into the conduction region.

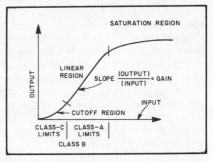

Fig. 7-27 — A typical transfer-function graph displaying output versus input for different amplifier classes.

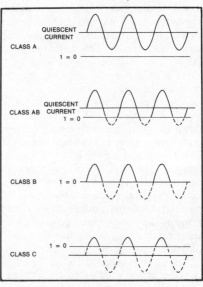

Fig. 7-28 — Amplifying-device output current for various classes of operation. All graphs assume a sine-wave input signal.

A class-C amplifier has a conduction angle of less than 180°. The output current will just be pulses at the signal frequency, as shown in Fig. 7-28D. The amplifier is cut off for considerably more than half the cycle, so the operating efficiency can be quite high — up to 80% with proper design. Linearity is very poor, however.

The linearity of the amplifier stage is important, because it describes how faithfully the input signal will be reproduced at the output. Any nonlinearity results in a distorted output. So you can see that a class-A amplifier will have the least amount of distortion, while a class-C amplifier produces a severely distorted output. A side effect of this distortion or nonlinearity is that the output will contain harmonics of the input signal. Odd-order intermodulation products are formed, which are close to the desired frequency, and so will not be filtered out by a resonant tank circuit. So a pure sine-wave input signal becomes a complex combination of sine waves at the output.

By now you are probably asking, "But why would anyone want to have an amplifier that generates a distorted output signal?" That certainly is a good question. At first thought, it sure doesn't seem like a very good idea. You must remember, however, that every circuit design consists of compromises between fundamentally opposing ideals. We would like to have perfect linearity for our amplifiers, but we would also like them to have 100% efficiency. You have just learned that those two ideals are exclusive. The closer you get to one of them, the further you get from the other. So you will have to compromise the ideals a bit to achieve a workable design. Your particular application and circuit conditions help you determine which compromises to make.

An important point to keep in mind is that most RF amplifiers will have a tuned-output tank circuit. That tuned circuit stores electrical energy like a mechanical flywheel, which is used to store mechanical energy. A heavy wheel, with its mass concentrated as close to the outer rim as possible, will have a large moment of inertia. That means that it has a lot of angular momentum when it is spinning, so it will continue to spin unless there is something to make it stop. If you turn off the motor that was used to set the flywheel in motion, the wheel will continue to spin for a long time. (In the ideal condition, where there is no friction, and the wheel is spinning on perfect bearings, it would never stop.) Your car engine uses a flywheel to store some of the energy from the gasoline exploding in the engine, and keep the system spinning smoothly. Without a flywheel, there would be a noticeable "bump" every time a cylinder fires.

A parallel tuned circuit is the electrical equivalent of a flywheel. (This type of circuit is often called a tank circuit because you can also think of it as a storage tank for electrical energy.) By placing such a circuit in series with the amplifier output, you can use it to smooth out the "bumps" that will occur when there is no output because the amplifier is turned off. This is especially useful if you are amplifying a pure sine-wave signal, such as for CW, and want to take advantage of the increased efficiency offered by a class-C amplifier.

The tank circuit is a parallel-resonant circuit, as discussed in Chapter 5. You should remember that the electric current flowing in the circuit is passing the energy back and forth between the electric field of the capacitor and the magnetic field of the inductor. The signal source only has to supply small amounts of energy to keep the current flowing. The tuned tank circuit will filter out the unwanted harmonics generated by a nonlinear amplifier stage. Usually, the tank circuit should have a Q of at least 20 to reduce these harmonics to an acceptable level.

If you are amplifying an audio signal, linearity may be the most important consideration. Use a class-A amplifier for audio stages. For an AM or SSB signal, which is an RF signal envelope that varies at an audio rate, you would want to use a linear amplifier. You may be willing to accept a bit of nonlinearity to obtain increased efficiency, in which case class-AB operation would be indicated. In some instances, you

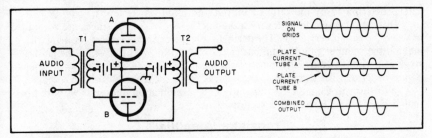

Fig. 7-29 — Two class-B amplifiers connected to operate as a push-pull amplifier. The graphs show current waveforms at the points indicated.

may even be willing to use a class-B amplifier, and let the flywheel effect of a tank circuit fill in the waveform voids for you.

You can take advantage of the nonlinearity of a class-C amplifier by using it as a frequency-multiplier stage. We mentioned earlier that one consequence of a nonlinear amplifier is that it would generate harmonic signals. If you want to multiply the frequency for operation on another band, sometimes you can do that by using a tank circuit tuned to a harmonic of the input frequency, and selecting the appropriate harmonic from the output of a class-C amplifier. This is especially useful for generating an FM signal at VHF or UHF, where you may start with a signal in the HF range, apply modulation through a reactance modulator, and then multiply the signal frequency to the desired range on 2 meters or higher. The exact bias point for the class-C amplifier will determine which harmonic frequency will be the strongest in the output, so careful selection of bias point is important.

Class-B amplifiers are often used for audio frequencies by connecting two of them back to back in push-pull fashion. Fig. 7-29 illustrates a simple triode-tube push-pull amplifier and the waveforms associated with this type of operation. A push-pull amplifier can also be used at RF. While one tube is cut off, the other is conducting, so both halves of the signal waveform are present in the output. This reduces the amount of distortion in the output, and will result in fewer harmonics.

Voltage, Current and Power Amplification

When we talk about amplifiers, it is common to think of a power amplifier. An amateur amplifier is said to be "1 W in, 10 W out," for example. We can also build circuits to amplify voltage or current, however. An instrumentation amplifier may be designed to amplify a very small voltage so that it can be measured with a voltmeter; in a multistage amplifier, we may wish to amplify the output current from one stage to drive the next stage. Stage, by the way, is the name we give to one of a number of signal-handling sections used one after another in an electronic device.

The input and output impedance of an amplifier circuit will vary, depending on what type of amplification the circuit is designed to provide. Generally speaking, voltage amplifiers have a very low imput impedance and a high output impedance. If too much current is drawn from the output of a voltage amplifier, its operation can be affected. Current and power amplifiers have a lower output impedance, and can be used to supply both voltage and current to a load.

Amplifier Gain

The gain of an amplifier is the ratio of the output signal to the input signal. Voltage amplifier gain is based on the ratio of output and input voltages, current amplifier gain on the ratio of output and input current levels and power amplifier gain is determined by the ratio of output and input power levels.

This ratio can get to be a very large number when several stages are combined or when the gain is very large. The decibel expresses the ratio in terms of a logarithm, making the number smaller and easier to work with. We often state the gain of a stage as a voltage gain of 16 or a power gain of 250, which are ratios, or refer to a gain of 24 dB, for example.

[Now turn to Chapter 10 and study FCC examination questions with numbers that begin 4AG-2. Review this section as needed.]

Transistor Amplifiers

We are limiting our discussion here to the operation of bipolar-transistor-amplifier circuits, but many of the techniques and general circuit configurations also apply to tube-type amplifiers. One major difference between tube-type amplifiers and transistor amplifiers is that tubes are voltage-operated devices, but transistors are current operated. A transistor amplifier is essentially a current amplifier. To use it as a voltage amplifier, the current is drawn through a resistor, and the resulting voltage drop provides the amplifier output. FET amplifiers operate as voltage amplifiers. To learn more about how tube-type amplifiers function, we recommend that you turn to appropriate sections of *The ARRL Handbook*. At some point in your Amateur Radio career, you will probably need to learn about tube amplifiers, but for the purpose of helping you pass your Advanced class exam, we will concentrate on transistor circuits.

Bipolar-transistor diode junctions must be forward biased in order to conduct significant current. If your circuit includes an NPN transistor, the collector and base must be positive with respect to the emitter, and the collector must be more positive than the base. When working with a PNP transistor, the base and collector must be negative with respect to the emitter and the collector more negative than the base. The required bias is provided by the collector-to-emitter voltage, and by the emitter-to-base voltage. These bias voltages cause two currents to flow: Emitter-to-collector current and emitter-to-base current. Either type of transistor, PNP or NPN, can be used with a negative- or positive-ground power supply. Forward bias must still be maintained, however. Remember that the amount of bias current sets the class of amplifier operation.

The lower the forward bias, the less collector current will flow. As the forward bias is increased, the collector current rises and the junction temperature rises. If the bias is continuously increased, the transistor eventually overloads and burns out. This condition is called thermal runaway. To prevent damage to the transistor, some form of bias stabilization should be included in a transistor amplifier design. Even if the bias is not increased, however, thermal runaway can occur. As the transistor heats up, its beta increases, causing more collector current to flow. This causes more heating, even higher beta and even more current, until eventually the transistor burns out.

Amplifier circuits used with bipolar transistors fall into one of three types, known as the common-base, common-emitter and common-collector circuits. These are shown in Fig. 7-30 in elementary form. The three circuits correspond approximately to the grounded-grid, grounded-cathode and cathode-follower vacuum-tube circuits, respectively.

Common-Emitter Circuit

The common-emitter circuit is shown in Fig. 7-30A. The base current is small and the input impedance is fairly high — several thousand ohms on average. The collector resistance is some tens of thousands of ohms, depending on the signal-circuit source impedance. The common-emitter circuit has a lower cutoff frequency than does the common-base-circuit, but it gives the highest power gain of the three configurations.

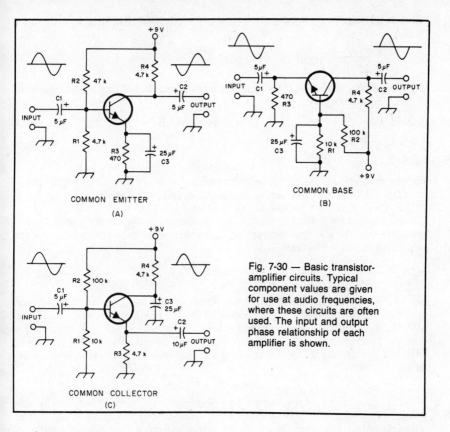

Fig. 7-30 — Basic transistor-
amplifier circuits. Typical
component values are given
for use at audio frequencies,
where these circuits are often
used. The input and output
phase relationship of each
amplifier is shown.

In this circuit, the output (collector) current phase is opposite to that of the input (base) current, so any feedback through the small emitter resistance is negative
and that will stabilize the amplifier, as we will show by an example.

Common-emitter amplifiers are probably the most common bipolar-transistor-
amplifer type, so we will use this circuit to illustrate some of the design procedures.
R1 and R2 make up a divider network to provide base bias. These resistors provide
a fixed, stable operating point, and tend to prevent thermal runaway. R3 provides
the proper value of bias voltage, when normal dc emitter current flows through it.
This biasing technique is often referred to as self bias. C3 is used to bypass the ac
signal current around the emitter resistor, and should have low reactance at the signal frequency. C1 and C2 are coupling capacitors, used to allow the desired signals
to pass into and out of the amplifier, while blocking the dc bias voltages. Their values
should be chosen to provide a low reactance at the signal frequency.

The emitter-to-base resistance is approximately:

$$R_{e-b} = \frac{26}{I_e}$$

(Eq. 7-8)

where I_e is the emitter current in milliamperes. The voltage gain is the ratio of collector load resistance to (internal) emitter resistance.

$$A_V = \frac{R_L}{R_{e-b}}$$

(Eq. 7-9)

For the example of Fig. 7-30A, if the emitter current is 1.6 mA, then R_{e-b} is 16.25 Ω, and

$$A_V = \frac{4.7 \text{ k}\Omega}{16.25 \text{ }\Omega} = 289$$

If you would like to express this voltage-gain ratio in decibels, just take the logarithm of 289, and multiply by 20:

Gain in dB = $20 \times \log(289) = 20 \times 2.46 = 49$ dB.

The base input impedance is given by:

$$R_b = \beta R_{e-b} \tag{Eq. 7-10}$$

For our example, $R_b = 1625 \text{ }\Omega$, assuming beta = 100. The actual amplifier input impedance is found by considering R1 and R2 to be in parallel with this resistance, so the input impedance for our amplifier is about 1177 Ω.

If you omit the emitter bypass capacitor, then all of the emitter signal current must flow through R3. This resistor dominates the emitter impedance in the voltage gain equation, so we have:

$$A_V = \frac{R_L}{R_E} \tag{Eq. 7-11}$$

Now we obtain a gain of 10 from our amplifier (20 dB). The base impedance of the unbypassed-emitter circuit becomes $\beta R_E = 47$ kΩ, which when combined with the bias resistors, gives an amplifier input impedance of 3.92 kΩ. Notice that the unbypassed emitter resistor introduced 29 dB of negative feedback. This has the effect of stabilizing the gain and impedance values over a wide frequency range.

Common-Base Circuit

The input circuit of a common-base amplifier must be designed for low impedance. Eq. 7-8 can be used to calculate the base-emitter junction resistance. The optimum output load impedance, R_L, may range from a few thousand ohms to 100,000 Ω, depending on the circuit. In the common-base circuit (Fig. 7-30B), the phase of the output (collector) current is the same as that of the input (emitter) current. The parts of these currents that flow through the base resistance are likewise in phase, so the circuit tends to be regenerative and will oscillate if the current amplification factor is greater than one. You will notice that the bias resistors have much the same configuration as with the common-emitter circuit, and that input- and output-coupling capacitors are used. C3 bypasses the base bias resistor, placing the base at ac-ground potential.

Common-Collector Circuit

The common-collector transistor amplifier, sometimes called an emitter-follower amplifier, has high input impedance and low output impedance. The input resistance depends on the load resistance, being approximately equal to the load resistance divided by $(1 - \alpha)$. The fact that input resistance is directly related to the load resistance is a disadvantage of this type of amplifier, especially if the load is one whose resistance or impedance varies with frequency.

The cutoff frequency of the common-collector circuit is the same as in the common-emitter amplifier. The input and output currents are in phase, as shown in Fig. 7-30C. This amplifier also uses input- and output-coupling capacitors, and a bypass capacitor on the collector. R3 is a feedback resistor, and is also used to develop an output voltage for the next stage.

[Study FCC examination questions with numbers that begin 4AG-10 and 4AG-11. Review this section as needed.]

RF Amplifiers

RF amplifiers are useful primarily to improve the receiver noise figure at frequencies around 28 MHz and above. The idea is to increase the level of weak RF signals above the internal noise of the receiver. An RF amplifier can degrade the receiver dynamic range, however, because the mixer will be overloaded more easily. If the mixer overloads, it is more likely to generate spurious mixing products, degenerating performance. When atmospheric and man-made noise levels exceed the noise generated in the mixer, an RF amplifier may not provide such a big advantage. Actually, it should be possible to realize better dynamic range by not using an RF amplifier. The gain of the RF stage, when one is used, should be set for the minimum level needed to override the mixer noise. Sometimes that is only a few decibels. A good low-noise device should be employed as the RF amplifier in such instances.

All of the FET amplifiers in Fig. 7-31 are capable of providing low-noise operation and good dynamic range. The common-source circuits at parts B and C can provide up to 25 dB of gain. However, they are more prone to instability than is the circuit at A. Therefore, the gates are shown tapped down on the gate tank, placing the input at a low impedance point on the tuned circuit to discourage self-oscillation. The same is true of the drain tap. JFETs are able to withstand higher RF voltage levels without damage than MOSFET devices can.

A broadband bipolar-transistor RF amplifier is shown in Fig. 7-31D. This type of amplifier will yield approximately 16 dB of gain up to 148 MHz, and it will be unconditionally stable because of the negative feedback in the emitter and base circuits. A broadband 4:1 transformer is used in the collector to provide a 200-ohm load on the transistor, and step the impedance down to approximately 50 ohms at the amplifier output. The input impedance to the 2N5179 is also about 50 ohms. A band-pass filter should be used at the input and output of the amplifier to provide selectivity.

RF Power Amplifiers

Most of our discussion here has centered around amplifiers used in receiver circuits. Of primary importance to most Amateur Radio operators is the goal of getting some RF energy to leave the station and travel to another location where it can be received — in other words, communication! Achieving that goal requires that the weak signals generated in an RF oscillator be amplified to some higher power level. Either tube or transistor amplifier circuits are used for that purpose.

Of course, we are always concerned about the efficiency of our circuits, but since a power amplifier is a larger drain on the power supply, it is a good idea to pay a little extra attention to that factor with a high-power amplifier. The goal is to transfer as much power to the load as possible. The total power generated by the amplifier is given by:

$$P_{IN} = P_{OUT} + P_D \qquad \text{(Eq. 7-12)}$$

where

P_{IN} is the dc- and drive-power input
P_{OUT} is the power delivered to the load
P_D is the power dissipated in the amplifier resistances

The efficiency is calculated from the input and output power:

$$\text{Eff.} = \frac{P_{OUT}}{P_{IN}} \times 100\% = \frac{P_{OUT}}{P_{OUT} + P_D} \times 100\% \qquad \text{(Eq. 7-13)}$$

where the efficiency is expressed as a percentage.

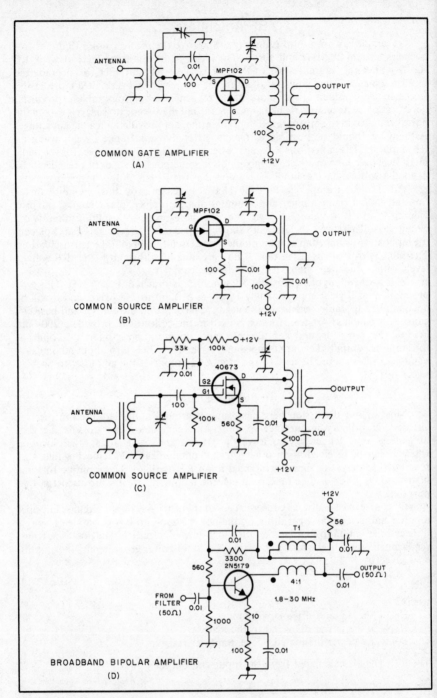

Fig. 7-31 — Narrowband FET RF amplifiers are shown at A, B and C. D is a broadband bipolar-transistor RF amplifier.

As an example, suppose the output from an amplifier is 1500 W, and there is a total of 500 W dissipated in the amplifier itself. Then the efficiency of that amplifier is:

$$\text{Eff.} = \frac{1500 \text{ W}}{1500 \text{ W} + 500 \text{ W}} \times 100\% = 75\%$$

The optimum load resistance for an amplifier is determined by the current transfer characteristics of the amplifying device. For a transistor amplifier, the optimum load resistance is:

$$R_L = \frac{V_{CC}^2}{2P_O} \qquad \text{(Eq. 7-14)}$$

where

V_{CC} is the dc collector voltage
P_O is the amplifier power output in watts

For example, if you have a transistor amplifier that uses a collector voltage of 25 V to produce an output power of 50 W, the load resistor should be:

$$R_L = \frac{(25 \text{ V})^2}{2 \times 50 \text{ W}} = \frac{625 \text{ V}^2}{100 \text{ W}} = 6.25$$

Vacuum tubes have more complex current-transfer characteristics than transistors, and selecting the optimum value of load resistor is a bit more difficult. The equation must include a term that varies with operating class. This also affects the maximum efficiency obtainable from each amplifier class. For a tube amplifier the optimum load-resistor value is given by:

$$R_L = \frac{V_P}{KI_P} \qquad \text{(Eq. 7-15)}$$

where

V_P is the dc plate potential in volts
I_P is the dc plate current in amperes
K is a constant that approximates the RMS current to dc current ratio appropriate for each operating class:

$$\text{class A,} \quad K \approx 1.3$$
$$\text{class AB,} \quad K \approx 1.5$$
$$\text{class B,} \quad K \approx 1.57$$
$$\text{class C,} \quad K \approx 2$$

What is the optimum value of load resistor for a class-C tube-type RF amplifier that has 1000-V dc applied to the plate and 500 mA of plate current flowing?

$$R_L = \frac{1000 \text{ V}}{2 \times 0.5 \text{ A}} = 1 \text{ k}\Omega$$

IF Amplifiers

As mentioned earlier, the main reason for using an intermediate frequency in a transmitter or receiver is that the circuits can process signals more efficiently if they are all converted to a single frequency first. Tuned circuits at the IF will not have to be readjusted for every small frequency change, and amplifiers can be designed for optimum operation at a single frequency. These amplifiers are designed to pass a small band of frequencies, instead of a wide range, and that means the circuit will have improved selectivity to block signals that are slightly off frequency.

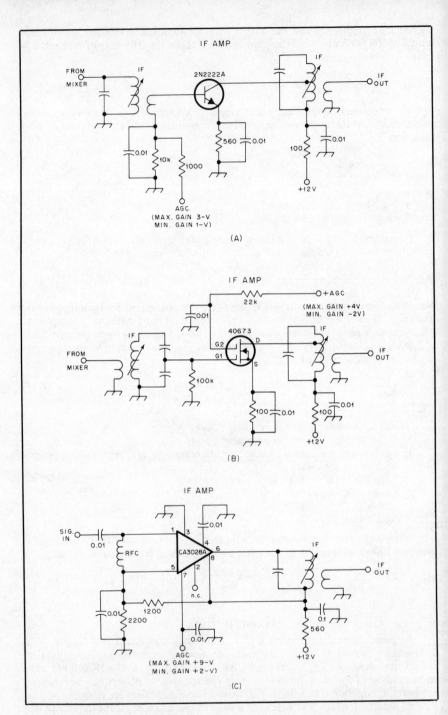

Fig. 7-32 — Typical IF amplifier stages are shown, including input for an AGC voltage. At A is a bipolar-transistor amplifier and B is a dual-gate MOSFET stage. An IC amplifier is shown at C.

The amount of amplification used in a receiver will depend on the signal level available at the input to the IF strip. Sufficient gain is needed to ensure ample audio output to drive headphones or a speaker. Another consideration is the amount of AGC-initiated IF-gain change. The more IF stages used (two is typical), the greater the gain change caused by AGC action. AGC stands for *automatic gain control*, and refers to a method of using a voltage derived from the input signal to control the stage gain. Stronger signals require less gain, and weaker ones require more. The problem is that if you have the audio gain turned all the way up to hear a weak signal and then a strong signal suddenly comes through, it will blast your eardrums. A good AGC system will quickly reduce the amplifier gain so the strong signal will not be amplified as much. The range is on the order of 80 to 120 dB of gain variation with AGC applied to a pair of typical IC amplifiers.

Nearly all modern receiver circuits use IC amplifiers in the IF section. Numerous types of ICs are available to provide linear RF and IF amplification at low cost. With careful layout techniques, IC amplifiers will be very stable. Bypass capacitors should be placed as close to the IC pins as possible. Input and output circuit elements must be separated to prevent mutual coupling, which can cause unstable operation. If IC sockets are used, they should be of the low-profile variety, with short socket conductors. Fig. 7-32 contains examples of bipolar transistor, FET, and IC IF amplifiers. Typical component values are given.

Intermediate Frequency Choice

Selection of an intermediate frequency is another compromise between two conflicting factors. With a lower IF, the amplifier gain will be higher, and the selectivity will be improved. But this means the image frequency is closer to the desired receive frequency, which reduces the image ratio. A higher IF improves the image ratio, but the gain and selectivity are reduced. It is a good idea to avoid selecting an IF on which there are strong signals, such as the broadcast band, because these signals may get into the IF amplifier and cause interference or other problems.

In a typical receiver IF system, the first amplifier stage is primarily responsible for providing selectivity. You want to limit the signals going through the IF section to only the ones you want to receive at that instant! Succeeding stages provide more gain and filtering. The final stage must provide an impedance match to the detector, and supply any final gain needed.

Amplifier Stability

Excessive gain or undesired feedback may cause amplifier instability. Oscillation may occur in unstable amplifiers under certain conditions. Damage to the active device from overdissipation is only the most obvious effect of oscillation. Deterioration of noise figure, spurious signals generated by the oscillation and reradiation of the oscillation through the antenna, causing RFI to other services, can also occur from amplifier instability. Negative feedback will stabilize an RF amplifier. Care in terminating both the amplifier input and output can produce stable results from an otherwise unstable amplifier. Attention to proper grounding and proper isolation of the input from the output by means of shielding can also yield stable operating conditions.

Neutralization

A certain amount of capacitance exists between the input and output circuits in any active device. In the bipolar transistor it is the capacitance between the collector and the base. In an FET it's the capacitance between the drain and gate. In vacuum tubes it is the capacitance between the plate and grid circuit. So far we have simply ignored the effect that this capacitance has on the amplifier operation. In fact, it doesn't have much effect at the lower frequencies. Above 10 MHz or so, however,

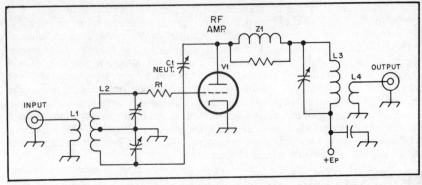

Fig. 7-33 — An example of a neutralization technique used with a tube-type RF amplifier.

the capacitive reactance may be low enough to cause complications. Oscillations can occur when some of the output signal is fed back in phase, so that it adds to the input (positive feedback). As the output voltage increases so will the feedback voltage: The circuit adds fuel to its own fire and the amplifier is now an oscillator. The output signal is no longer dependent on the input signal, and the circuit is useless as an amplifier.

In order to rid the amplifier of this positive feedback, it is necessary to provide a second feedback path, which will supply a signal that is 180° out of phase with the positive feedback voltage. This is called negative feedback. This path should supply a voltage that is equal to that causing the oscillation, but of opposite polarity.

One *neutralization* technique for vacuum-tube amplifiers is shown in Fig. 7-33. In this circuit, the neutralization capacitor is adjusted to have the same value as the interelectrode capacitance that is causing the oscillation.

With solid-state amplifiers, a similar technique could be used, although the interelement capacitances tend to be much smaller. It is more common to include a small value of resistance in either the base or collector lead of a low-power amplifier. Values between 10 and 20 ohms are typical. For higher power levels (above about 0.5 W), one or two ferrite beads are often used on the base or collector leads.

Parasitic Oscillations

Oscillations can occur in an amplifier on frequencies that have no relation to those intended to be amplified. Oscillations of this sort are called *parasitics* mainly because they absorb power from the circuits in which they occur. Parasitics are brought on by resonances that exist in either the input or output circuits. They can also be below the operating frequency, which is usually the result of an improper choice of RF chokes and bypass capacitors. High-Q RF chokes should be avoided, because they are most likely to cause a problem.

Parasitics are more likely to occur above the operating frequency as a result of stray capacitance and lead inductance along with interelectrode capacitances. In some cases it is possible to eliminate such oscillations by changing lead lengths or the position of leads so as to change the capacitance and inductance values. An effective method with vacuum tubes is to insert a parallel combination of a small coil and a resistor in series with the grid or plate lead. The coil serves to couple the VHF energy to the resistor, and the resistor value is chosen so that it loads the VHF circuit so heavily that the oscillation is prevented. Values for the coil and resistor have to be found experimentally as each different layout will probably require different suppressor networks.

With transistor circuits, ferrite beads are often used on the device leads, or on a short connecting wire placed near the transistor. These beads act as a high impedance to the VHF or UHF oscillation, and block the parasitic current flow. In general, proper neutralization will help prevent parasitic oscillations.

Transmitter Controls

Most of the newer solid-state transceivers have broadband, no-tune final amplifiers. Many popular rigs include a tube-type final amplifier, and most external power amplifiers used to increase your signal beyond the typical 100-W level, rely on vacuum tubes. These usually have a pi-network output circuit that must be adjusted carefully for proper opera-

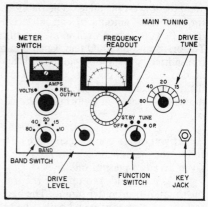

Fig. 7-34 — Typical controls found on simple transmitters with a tube-type final amplifier.

tion. You should be familiar with the manufacturer's operating instructions, or the function of each control on a piece of homemade gear. It is a good idea to know the general tune-up procedure for such equipment, however.

Fig. 7-34 shows the front-panel controls of a simple CW transmitter, which will serve to illustrate the tune-up technique. The BAND switch selects the tuned circuits for the various stages in the transmitter. Each band-switch position selects the proper filters, and switches the tap point on the tank-circuit inductor, so that each circuit can be tuned to the operating frequency. The tuning and loading controls are associated with the final output stage. These are two variable capacitors in the output pi network. The tuning control is a capacitor connected from the collector or plate of the amplifying device to ground. The loading control is a variable capacitor connected across the output.

If your transmitter has a control labelled DRIVE TUNE it more than likely adjusts a variable element in the tuned circuits of the low-level stages. This ensures that all of the stages ahead of the final amplifier are tuned to resonance, and maximum power is fed to the output stage. A DRIVE LEVEL control is used to adjust the amount of power that is fed to the output stage for amplification. Since it is possible to over-drive the output stage of most transmitters (there is usually a surplus of drive available), this control should be set for the proper amount of drive in accordance with the tube or equipment manufacturer's specifications. Nothing is gained by overdriving an amplifier stage. The output device may be driven beyond its ratings, in which case shortened device life or destruction may occur. Chances are that increased amounts of harmonic energy and other spurious undesired output will be transmitted along with the main signal. This will increase the chances of interference to other radio services and increase the possibility of television interference. The best motto is to keep the signal clean, and this can only be done by running the amplifying devices within their specifications.

Meters and what they measure will vary from one rig to another. Often there will be a switch to change the meter function, and with a transceiver there is usually an automatic change in what the meter indicates when you switch from receive to transmit. However the meter is wired, the main goal is to allow to you tune the final amplifier properly. The meter may have a VOLTS position, which would indicate the supply voltage to the output stage. There may also be a position for CURRENT, which would be a measure of the amount of current drawn by the final stage. Additonally there may or may not be a position for relative RF output. This position is helpful for tuning the output stage by adjusting the tuning and loading controls for a maximum reading on the meter. The DRIVE TUNE and DRIVE LEVEL controls are adjusted for a specified amount of current, according to the manufacturer's suggestions.

The FUNCTION switch in most transmitters will have at least several positions. The OFF position should remove all power from the transmitter. STANDBY places the transmitter in a standby mode, with no output, but everything ready to go when needed. The TUNE position allows the operator to make preliminary adjustments of the transmitter at reduced power levels, and the OPERATE position allows the transmitter to function in a normal fashion.

Tuning Procedure

If you own a commercial piece of transmitting equipment, it is always best to follow the manufacturer's instructions on how to tune the transmitter. The outline given here is a general one at best, but, will give you an idea of how it's done.

It is a good idea to operate the transmitter into a dummy load while learning the ropes of transmitter tuning. The best dummy load is a high-power noninductive resistor submerged in an oil bath. The value of resistance used is normally 50 ohms, the same as the output impedance of most tramsmitters. Turn the FUNCTION switch to STANDBY. A warm-up period of a minute or so should be sufficient to allow the tube filaments to reach operating temperature. The band switch must be placed in the proper position for the intended band of operation.

Set the meter switch to the GRID or RELATIVE OUTPUT position, turn the FUNCTION switch to TUNE and key the transmitter. With the DRIVE LEVEL advanced to the approximate mid-rotation position, adjust the DRIVE TUNE control for a maximum reading. Let up on the key and switch the meter to the CURRENT or PLATE position, and turn the FUNCTION switch to the OPERATE position. With the LOAD control set for maximum capacitance, again key the transmitter and quickly adjust the TUNE control for a dip in plate current. Adjust the final-amplifier LOAD control for a peak current reading, then redip the plate current with the TUNE control. Continue this procedure through several iterations, until the tube is drawing the specified amount of current. It is a good idea to check the grid current after this procedure, since it has likely changed during tuning. It should be readjusted to the manufacturer's specification by means of the drive control. Never run the transmitter outside the ratings specified by the manufacturer.

[Before proceeding, turn to Chapter 10 and study FCC examination questions with numbers that begin 4AG-7 and 4AG-9. Review this section as needed.]

OSCILLATORS

In the previous section we described an amplifier-circuit problem that occurs if some of the output signal makes its way back to the input in a manner that creates positive feedback. This circuit instability turns the amplifier into an oscillator. When we want the circuit to behave as an amplifier, this is a problem. But sometimes we want a circuit that will generate a signal (often at radio frequencies) without any input signal, and in that case we can take advantage of the instability of an amplifier.

To start a circuit oscillating, we need to feed power from the plate back to the grid of a tube or from the collector to the base of a transistor. This is called feedback. Feedback is what happens when you are using a public address system and you get the microphone too close to the loudspeaker. When you speak, your voice is amplified in the public address system and comes out over the loudspeaker. If the sound that leaves the speaker enters the microphone and goes through the whole process again, the amplifier begins to squeal, and the sound keeps getting louder until the "circuit" is broken, even though you are not talking into the microphone anymore. Usually, the microphone has to be moved away from the speaker or the volume must be turned down. This is a good example of an amplifier becoming unstable through positive feedback, and breaking into oscillation.

When the output signal applied to the input reinforces the signal at the input, we call it positive feedback. Negative feedback opposes the regular input signal.

Positive feedback can be created in many ways — so many ways, in fact, that we can't possibly cover them all here. There are three major oscillator circuits used in Amateur Radio. They are shown in Fig. 7-35. These oscillators can be built using vacuum tubes or FETs, and the feedback circuits will be much the same as the ones shown here for bipolar transistor circuits. The amount of feedback required to make the circuit oscillate is determined by the circuit losses. There must be at least as much energy fed back to the input as is lost in heating the components, or the oscillations will die out.

The Hartley oscillator uses inductive feedback. Alternating current flowing through the lower part of the tapped coil induces a voltage in the upper part, which is connected to the input of the circuit. A Hartley oscillator is the least stable of the three major oscillator types.

The second general type is the Colpitts oscillator circuit. It uses capacitive feedback. The collector-circuit energy is fed back by introducing it across a capacitive voltage divider, which is part of the tuning-circuit capacitance. This coupling sets up an RF voltage across the whole circuit, and is consequently applied to the transistor base. The large values of capacitance used in the tuning circuit tend to stabilize the circuit, but the oscillation frequency still will not be as stable as a crystal oscillator.

The most stable oscillator circuit is the Pierce crystal oscillator. The Pierce circuit uses capacitive feedback, with the necessary capacitances supplied by C10 and C11. For a tube-type Pierce oscil-

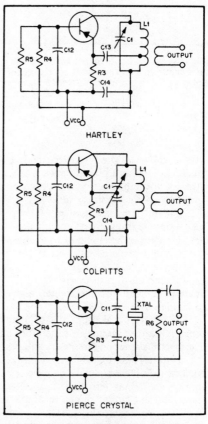

Fig. 7-35 — Three common types of transistor oscillator circuits. Similar circuits can be built using tubes or FETs.

lator, the feedback capacitance is supplied by the tube interelectrode capacitances. Besides its stability, another reason why the Pierce circuit is popular is that you do not have to build and tune an LC tank circuit. Adjusting the tank circuit for the exact resonant frequency can be touchy, but it is a simple matter to plug a crystal into a circuit and know it will oscillate at the desired frequency!

Quartz Crystals

A number of crystalline substances can be found in nature. Some have the ability to change mechanical energy into an electrical potential, and vice versa. This prop-

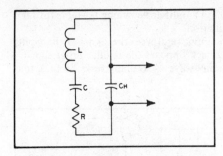

Fig. 7-36 — The electrical equivalent circuit of a quartz crystal. L, C and R are the electrical equivalents of the crystal mechanical properties and C_H is the capacitance of the holder plates, with the crystal serving as the dielectric.

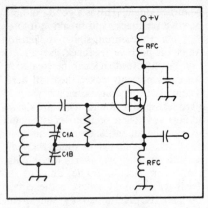

Fig. 7-37 — A Colpitts VFO circuit.

erty is known as the piezoelectric effect. A small plate or bar properly cut from a quartz crystal and placed between two conducting electrodes will be mechanically strained when the electrodes are connected to a voltage source. The opposite can happen, too. If the crystal is squeezed, a voltage will develop between the electrodes.

Crystals are used in microphones and phonograph pickups, where mechanical vibrations are transformed into alternating voltages of corresponding frequency. They are also used in headphones to change electrical energy into mechanical vibration.

Crystalline plates have natural frequencies of vibration ranging from a few thousand hertz to tens of megahertz. The vibration frequency depends on the kind of crystal, and the dimensions of the plate. What makes the crystal resonator (vibrator) valuable is that it has an extremely high Q, ranging from a minimum of about 20,000 to as high as 1,000,000.

The mechanical properties of a crystal are very similar to the electrical properties of a tuned circuit. We therefore have an "equivalent circuit" for the crystal. The electrical coupling to the crystal is through the holder plates, which "sandwich" the crystal. These plates form, with the crystal as the dielectric, a small capacitor constructed of two plates with a dielectric between them. The crystal itself is equivalent to a series-resonant circuit and, together with the capacitance of the holder, forms the equivalent circuit shown in Fig. 7-36.

Can we change the crystal in some way so that it will resonate at a different frequency? Sure. If we cut a new crystal, and make it longer or thicker, the resonant frequency will go down. On the other hand, if we want the crystal to vibrate at a higher frequency, we would make it thinner and shorter.

There are two major limitations to the use of crystals. First, we can't have more than two terminals to the circuit, since there are only two crystal electrodes. In other words, a crystal can't be tapped as we might tap a coil in a circuit. Second, the crystal is an open circuit for direct current, so you can't feed operating voltages through the crystal to the circuit.

Advantages and Disadvantages of Crystals

The major advantage of a crystal used in an oscillator circuit is its frequency stability, especially with mechanical vibration. The spacing of coil turns can change with vibration, and the plates of a variable capacitor can move. On the other hand, the frequency of a crystal is much less apt to change when the equipment is bounced around (short of actually cracking the crystal, anyway).

A crystal used in an oscillator circuit is easily affected by temperature, although it is affected much less than for some other oscillator circuits. Crystals are sometimes identified according to what plane of the crystal structure is cut to produce the element. These are labeled X-, Y- or Z-cut crystals. Often the crystal faces are cut at some oblique angle to the major axes. By careful selection of the type of cut, manufacturers are able to control the modes of vibration, temperature coefficient and other parameters. An X cut gives a crystal with a negative temperature coeffieient, which means that when the temperature of the crystal rises, the vibration frequency decreases. Other crystals are cut so they have positive temperature coefficients, and when their temperature rises, their oscillation frequency also rises.

Variable-Frequency Oscillators

The major advantage of a variable-frequency oscillator (VFO) is that a single oscillator can be tuned over a wide frequency range. Since the frequency is determined by the coil and capacitors in the tuned circuit, the frequency is often not as stable as in a crystal-controlled circuit. The Colpitts oscillator (Fig. 7-37) is one of the more common types of VFO. The FET in the diagram is connected like a source follower (similar to the bipolar-transistor emitter follower) with the input (gate) connected to the top of the tuned circuit and the output (source) tapped down on the tuned circuit with a capacitive divider. Although the voltage gain of a source follower can never be greater than one, the voltage step-up that results from tapping down on the tuned circuit ensures that the signal at the gate will be large enough to sustain oscillation. Even though the FET voltage gain is less than one, the power gain must be great enough to overcome losses in the tank circuit and the load.

The amplitude and frequency stability of a variable-frequency Colpitts oscillator is quite good, although it doesn't approach that of a crystal-controlled oscillator. The large capacitances used in the tank circuit minimize frequency shifts that are caused by small capacitance variations caused by vibration or variations in tube or transistor characteristics.

Most oscillators provide an output rich in harmonic content. The relative amplitude of any given harmonic can be enhanced by choosing the optimum bias voltage. A Colpitts oscillator that has a second tank circuit tuned to the desired harmonic and connected to the output makes an excellent harmonic generator.

FREQUENCY SYNTHESIZERS

Frequency synthesizers serve much the same purpose as VFOs. That is, they are used to provide a stable, variable tuning range, generally to control the operating

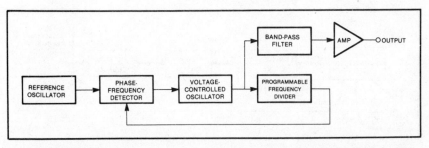

Fig. 7-38 — Block diagram of an indirect frequency synthesizer, which uses a phase-locked loop and a variable-ratio divider.

frequency of a radio. They are much more stable with changes in temperature and vibration than are Hartley or Colpitts VFOs. There are two major methods of frequency synthesis, direct and indirect synthesis.

Direct synthesis is a seldom-used technique that employs a series of switch-selectable oscillators and several mixers to change the frequency coverage of the final output. Sometimes one or more frequency-divider sections are included to add another dimension of control over the final output.

Indirect synthesizers are the type most commonly used in modern equipment. These use some specialized integrated-circuit chips in a phase-locked-loop (PLL) configuration. See Fig. 7-38. The signal frequency from a voltage-controlled oscillator (VCO) is divided by some integer value in a programmable-divider IC. The divider output is compared with a stable reference frequency in a phase-detector circuit. The phase detector produces an output that indicates the phase difference between the VCO and the reference signal. This signal is then fed back to the VCO through a filter. The loop provides a feedback circuit that tends to adjust the phase-detector output to zero. That means the divider output is always on the same frequency as the reference oscillator. By changing the division factor, you can change the synthesizer output. Mathematically:

$$F = NF_r \qquad \text{(Eq. 7-16)}$$

where F is the synthesizer output frequency, N is the division factor and F_r is the reference-oscillator frequency.

[Study FCC examination questions with numbers that begin 4AG-5. Review this section as needed.]

ANTENNA COUPLING

An antenna-coupling circuit has two basic purposes: (1) to match the output impedance of a power-amplifier tube or transistor to the input impedance of the antenna feed line, so the amplifier has a proper resistive load, and (2) to reduce unwanted emissions (mainly harmonics) to a very low value.

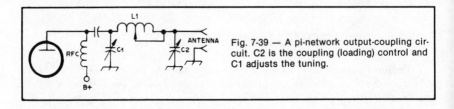

Fig. 7-39 — A pi-network output-coupling circuit. C2 is the coupling (loading) control and C1 adjusts the tuning.

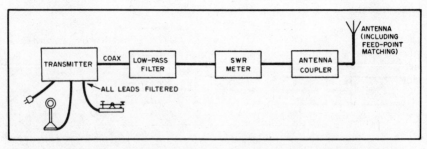

Fig. 7-40 — Proper connecting arrangement for a low-pass filter and antenna coupler. The SWR meter is included for use as a tuning indicator for the antenna coupler.

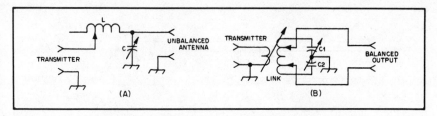

Fig. 7-41 — An L-network antenna coupler, useful for an unbalanced feed system is shown at A. B shows an inductively coupled circuit for use with a balanced feed line.

Most tube-type transmitters or amplifiers use *pi-network output-coupling* circuits (Fig. 7-39). The circuit is called a pi network because it resembles the Greek letter pi (π) — if you use your imagination a bit while you look at the two capacitors drawn down from ends of the horizontally drawn inductor. Because of the series coil and parallel capacitors, this circuit acts as a low-pass filter to reduce harmonics, as well as acting as an impedance-matching device. (A pi network with two coils shunted to ground and a series capacitor would make a high-pass filter.) The circuit Q will be equal to the plate load impedance divided by the reactance of C1. Coupling is adjusted by varying C2, which generally has a reactance somewhat less than the load resistance (usually 50 ohms). Circuit design information for pi networks appears in *The ARRL Handbook*.

Harmonic radiation can be reduced to any desired level by sufficient shielding of the transmitter, filtering of all external power and control leads, and inclusion of a low-pass filter (of the proper cutoff frequency) connected with shielded cable to the transmitter antenna terminals (see Fig. 7-40). Unfortunately, low-pass filters must be operated into a load of close to their design impedance or their filtering properties will be impaired, and damage may occur to the filter if high power is used. For this reason, if the filter load impedance is not within limits, a device must be used to transform the load impedance of the antenna system (as seen at the transmitter end of the feed line) into the proper value. For this discussion we will assume this impedance to be 50 ohms.

Impedance-matching devices are variously referred to as Transmatches, matchboxes or antenna couplers. Whatever you call it, an impedance-matching unit transforms one impedance to be equivalent to another. To accomplish this it must be able to cancel reactances (provide an equal-magnitude reactance of the opposite type) and change the value of the resistive part of a complex impedance.

The *L network* (Fig. 7-41A) will match any unbalanced load with a series resistance higher than 50 ohms. (At least it will if you have an unlimited choice of values for L and C.) Most unbalanced antennas will have an impedance that can be matched with an L network. To adjust this L network for a proper match, the coil tap is moved one turn at a time, each time adjusting C for lowest SWR. Eventually a combination should be found that will give an acceptable SWR value. If, however, no combination of L and C is available to perform the proper impedance transformation, the network may be reversed input-to-output by moving the capacitor to the transmitter side of the coil.

The major limitation of an L network is that a combination of inductor and capacitor is normally chosen to operate on only one frequency band because a given LC combination has a relatively small impedance-matching range. If the operating frequency varies too greatly, a different set of components will be needed.

If you are using balanced feed lines to your antenna, they may be tuned by means of the circuit shown in Fig. 7-41B. A capacitor may be added in series with the input

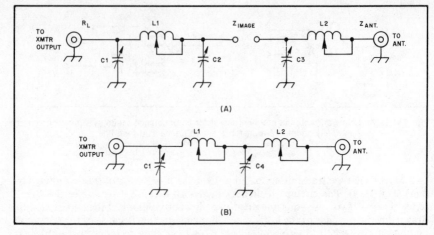

Fig. 7-42 — The pi-L network uses a pi network to transform the transmitter output impedance (R_L) to the image impedance (Z_{IMAGE}). An L network transforms Z_{IMAGE} to the antenna impedance, Z_{ANT}.

to tune out link inductance. As with the L network, the coil taps and tuning-capacitor settings are adjusted for lowest SWR, with higher impedance loads being tapped farther out from the coil center. For very low load impedances, it may be necessary to put C1 and C2 in series with the antenna leads (with the coil taps at the extreme ends of the coil).

You can convert an L network into a pi network by adding a variable capacitor to the transmitter side of the coil. Using this circuit, any value of load impedance (greater or less than 50 ohms) can be matched using some values of inductance and capacitance, so it provides a greater impedance-transformation range. Harmonic suppression with a pi network depends on the impedance-transformation ratio and the circuit Q.

If you need more attenuation of the harmonics from your transmitter, you can add an L network in series with a pi network, to build a pi-L network. Fig. 7-42A shows a pi network and an L network connected in series. It is common to include the value of C2 and C3 in one variable capacitor, as shown at Fig. 7-42B. The pi-L network provides the greatest harmonic attenuation of the three most-used matching networks.

[Turn to Chapter 10 and study FCC examination questions with numbers that begin 4AG-3. Review this section as needed.]

FILTERS: HIGH PASS, LOW PASS AND BAND PASS

The function of a filter is to transmit a desired band of frequencies without attenuation and to block all other frequencies. The resonant circuits discussed in previous sections all do this. However, the term "filter" is frequently reserved for those networks that transmit a desired band with little variation in output, and in which the transition from the "pass" band to the "stop" band is very sharp, rather than being a gradual change, as is the case with simple resonant circuits.

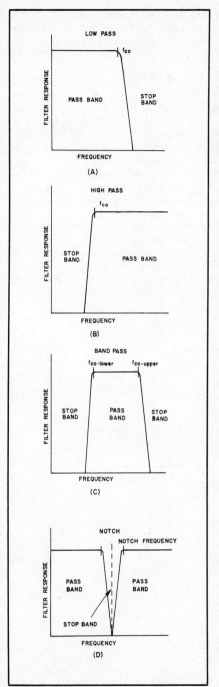

Fig. 7-43 — Ideal filter-response curves for low-pass, high-pass, band-pass and notch filters.

Filter Classification

Filters are classified into two general groups. A low-pass filter is one in which all frequencies below a specified frequency (called the cutoff frequency) are passed without attenuation. Above the cutoff frequency, the attenuation changes with frequency in a way that is determined by the network design. Usually you want this transition region to be as sharp as possible. A high-pass filter is just the opposite; there is no attenuation above the cutoff frequency, but attenuation does occur below that point.

High- and low-pass filters can be combined to make two more filter types. With these filters there are two cutoff frequencies, an upper and a lower one. With a band-pass filter, frequencies on both sides of the passband are attenuated. A pair of filters, one high-pass and one low-pass type, tuned to have overlapping passbands, make a simple band-pass circuit. If the cutoff frequencies of the high- and

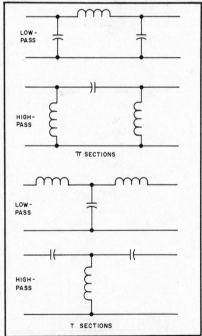

Fig. 7-44 — High- and low-pass pi and T filter networks are shown.

low-pass filters are brought close together, but not overlapping, you have a notch filter. In this case, signals on either side of a certain band are passed, while signals in the middle are attenuated. Fig. 7-43 illustrates the response curves for the four types of ideal filters.

Pi- and T-Network Filter Sections

A filter section can be either a π- or T-type network in which the series and shunt reactances are of opposite types, as shown in Fig. 7-44. Unlike the pi networks described in the section on antenna couplers, these sections are not designed for transforming a given value of load resistance into a different value that will be suitable for the source of power.Instead, the design is such that the output impedances will be reflected to the input, for frequencies in the passband.

For this to occur, the values of L and C in the section must have a characteristic, or image, impedance (in the passband) that equals the load resistance. Then power is transferred from the source to the load without attenuation. Another consequence of having the input and output impedances the same is that sections may be cascaded, or strung together one after another, with no change in the input impedance shown to the source of power, at least at the desired pass frequencies.

The advantage to cascading filter sections is that in the stopband — the range of frequencies in which attenuation occurs — the attenuation for signals increases with the number of filter sections, while remaining zero for signals in the passband. If the attenuation of a section at a given frequency is expressed in decibels, each similar

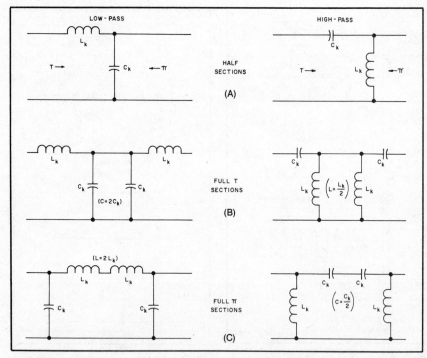

Fig. 7-45 — Diagram showing how filter half sections can be combined to form full pi and T filter sections.

section that is added to a filter also adds the same number of decibels to the overall attenuation at the same frequency.

The basis for filter design is the *half section*, shown in Fig. 7-45A. The inductance and capacitance, usually designated L_k and C_k, in the half section are related to a desired value of resistance, R, which is the termination or load impedance for the filter. Looking into the half section from the series end shows the beginning of a T section, and looking into it from the opposite side, the shunt end, shows the beginning of a pi section. The full sections are formed by connecting two half sections together as shown in Fig. 7-45, parts B and C. When two like reactances are in parallel, as in the shunt arm of the full T sections, they are usually combined into a single physical element. Similarly, where two are in series they may be combined into a single element.

Pay special attention to the way the elements are arranged in these sections. A low-pass filter uses an inductor in series with the signal and a capacitor shunted to ground. The inductor has little reactance at low frequencies, and increasing reactance at higher frequencies, so it will tend to block the high-frequency signals more. The capacitor will shunt any high-frequency energy to ground, but will look like a high impedance to any low-frequency energy. Likewise, a high-pass filter uses a capacitor in series with the signal to pass high frequencies and block low frequencies. An inductor shunted to ground will take any remaining low-frequency energy to ground, but will act as a high impedance to high-frequency signals. You should be able to determine if a filter section is a high-pass or a low-pass one by looking at the element arrangement.

Image Impedance

Filter components may be selected to build a filter of specified design by using the *image-parameter technique*. This method requires values to be calculated using rather complicated equations, which you do not need to know for your exam. The technique is based on the fact that the image impedance (the load impedance reflected back to the input) of a given filter section is not constant at all frequencies in the passband, nor are the image impedances alike for pi and T sections using the same L_k and C_k. The terminating impedance, R, is therefore a "nominal" value.

Because the filter input impedance has some variation, there will be an impedance mismatch at the input, and this results in a signal loss at some frequencies. This mismatch loss is entirely between the source and the filter input. There is no loss of power in the filter itself. Since the filter contains only pure reactances, any power that enters it is either delivered to the terminating resistance R or returned to the source. If there is an adjustable impedance-matching network between the power source and the filter, the mismatch loss can be overcome. This is usually the case when the filter follows a transmitting power amplifier.

Attenuation in the Stopband

In the stopband, the impedances are deliberately mismatched to prevent transfer of power to the load. In this band, with the type of filter sections considered so far, the attenuation increases progressively as the applied frequency is moved away from the cutoff frequency.

Note that the transition from the passband to the stopband is rather abrupt. With an actual filter section there will be variations in attenuation in the stopband near the cutoff frequency, and the sharp transition will be smoothed off. At frequencies somewhat removed from the cutoff point, the real curve becomes free from these variations and approaches the theoretical curve. Beyond about twice the cutoff frequency in the low-pass case, and about half the cutoff frequency in the high-pass case, the attenuation increases uniformly at 12 dB per octave. (An octave is equal to a frequency ratio of 2 to 1.)

Again there is no loss of power in the filter itself. The loss is because of a mis-

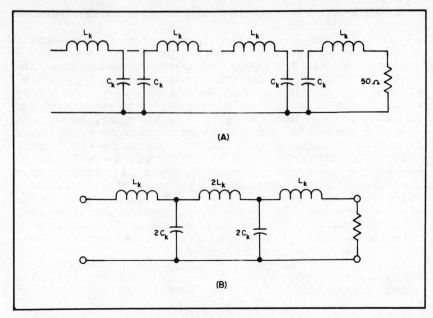

Fig. 7-46 — Half sections are cascaded to build a constant-k filter.

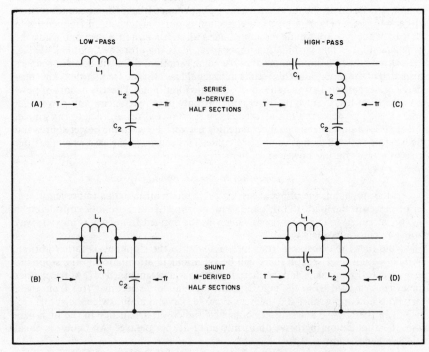

Fig. 7-47 — Schematic diagrams for series and shunt m-derived half sections.

match between the power source and the filter input impedance; in other words, the filter simply refuses to accept power from the source.

Constant-K Filter Design

Filter sections of the type considered so far are known as constant-k sections. *Constant-k filters* are easily built up from the basic sections shown in Fig. 7-46A. The values of the components can be found by calculation or by using reactance charts. Since each series filter element and each shunt element is either a single capacitor or a single inductor, the product of the reactances for a single series and a single shunt element is a constant for all frequencies.

The farther from the cutoff frequency a given signal is in the stopband, the more it will be attenuated. A constant-k filter never actually reaches a point of infinite attenuation. Constant-k filters are suitable for suppressing the harmonic output of a transmitter operating well below the filter cutoff frequency, but the closer the harmonic is to the cutoff frequency, the less it will be attenuated. Under typical operating conditions, you could expect the second harmonic to be attenuated about 30 dB. You can increase the stop-band attenuation by cascading several filter sections. Fig. 7-46 shows an example of a filter built by cascading four half sections.

M-Derived Half Sections

If you need additional attenuation at some particular frequency in the stopband that is close to the cutoff frequency, *m-derived* half sections or full sections can be used instead of the constant-k type. (You have probably guessed that m is an image-parameter constant.) There are two general types of m-derived sections. In the series m-derived type, additional reactance is introduced in series with the shunt element to form a circuit that is series-resonant at the frequency to be suppressed. This short-circuits the output at that frequency, and (theoretically) the attenuation is infinite. In the shunt m-derived filter the series arm contains a parallel-resonant circuit at the frequency to be suppressed. Such a circuit theoretically has an infinite impedance, so the undesired frequency is prevented from reaching the output end of the filter.

The basic low- and high-pass m-derived sections are shown in Fig. 7-47. Full sections, as well as multisection filters, can be formed by cascading sections as we discussed before. M-derived filters are usually designed to be used between two constant-k sections

Modern Network Design

Image-parameter filter designs are inexact approximations for actual filter frequency responses. So aside from the limitations of impedance variation and steepness of the cutoff-response curve, the calculations will give results that have limited accuracy. These designs have been largely surpassed by newer techniques that are based on exact mathematical equations that can be applied to filter characteristics. Some examples are filters based on equations called Butterworth and Chebyshev polynomials and elliptical functions. With these mathematical techniques, it is possible to build a catalog of filter characteristics, with appropriate component values, and to select a design from the tabulated data. Tables summarizing these computations can be found in *The ARRL Handbook*, and other reference books.

There are many kinds of so-called "modern filters," and they are usually referred to by the name of the mathematical function used to calculate the design. *Butterworth, Chebyshev and elliptical filters* are three kinds that have many applications in Amateur Radio. A Butterworth filter is used when you want a response that is as flat as possible in the passband, with no ripple. (Ripple is a variation of attenuation, and you could get these "ups and downs" inside the passband and/or outside the passband.) Unfortunately, the transition from passband to stopband is not very sharp with a Butterworth filter. The Chebyshev filter has a sharp cutoff, with some

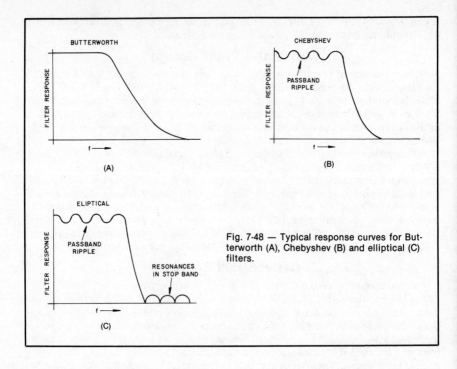

Fig. 7-48 — Typical response curves for Butterworth (A), Chebyshev (B) and elliptical (C) filters.

ripple in the passband. Higher SWR increases the ripple. The elliptical filter has the sharpest cutoff, with ripple in the passband and stopband. The elliptical filter also has infinite-rejection notches in the stopband, which can be positioned at specific frequencies that you want to attenuate. Fig. 7-48 illustrates the filter-response curves for these three types.

Mechanical Filters

The filters we have been talking about so far all use electronic circuits to provide the filtering action. The IF section of a radio requires very good band-pass filters to provide the narrow bandwidth needed for a top-performance rig. Two types of filters that do not use inductors and capacitors as the primary circuit elements are often used in the IF section of a radio. In Chapter 6, we described crystal-lattice filters and their operation. Filters of this type, with piezoelectric quartz crystals to provide

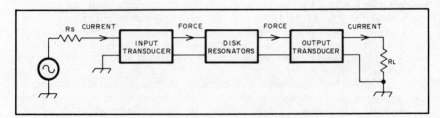

Fig. 7-49 — Block diagram of a mechanical filter.

high-Q, narrow-bandwidth characteristics are often used in modern receivers and transmitters.

A second type of non-electrical filter that is sometimes used in the IF stage of a radio is the *mechanical filter*. This filter uses a series of mechanical disks, which have be cut to size so they have a natural physical resonance at the desired frequency. Fig. 7-49 is a block diagram of mechanical-filter action. The filter has an input transducer to change the electrical signal into a mechanical force that is applied to the disks to make them vibrate. The tendency is for the disks to vibrate at their natural resonant frequency, but not at other frequencies applied to them. When the vibrations get to the output disk, it applies a force on the output transducer, creating an electrical signal from the mechanical vibrations. The undesired frequencies have been removed, to a great extent.

Mechanical filters can be made to operate in the range of about 60 kHz to 600 kHz. Below that the disks become unreasonably large and above it they become too small to be practical. Bandwidths of as little as 0.1% of the center frequency are possible. This means that a filter with a 455 kHz center frequency could have a bandwidth of 45.5 Hz!

[Before going on to Chapter 8, turn to Chapter 10 and study FCC examination questions with numbers that begin 4AG-4. Review this section as needed.]

Key Words

Amplitude modulation — A method of superimposing an information signal on an RF carrier wave in which the amplitude of the RF envelope (carrier and sidebands) is varied in relation to the information signal strength.

Circular polarization — Describes an electromagnetic wave in which the electric and magnetic fields are rotating. If the electric field vector is rotating in a clockwise sense, then it is called right-hand polarization and if the electric field vector is rotating in a counterclockwise sense, it is called left-hand polarization.

Deviation — The peak difference between an instantaneous frequency of the modulated wave and the unmodulated-carrier frequency in an FM system.

Deviation ratio — The ratio of the maximum frequency deviation to the maximum modulating frequency in an FM system.

Dielectric constant — A property of insulating materials that serves as a measure of how much electric charge can be stored in the material with a given voltage.

Dielectric materials — Materials in which it is possible to maintain an electric field with little or no additional energy being supplied. Insulating materials or nonconductors.

Electric field — A region through which an electric force will act on an electrically charged object.

Electric force — A push or pull exerted through space by one electrically charged object on another.

Electromagnetic waves — A disturbance moving through space or materials in the form of changing electric and magnetic fields.

Facsimile — The process of scanning pictures or images and converting the information into signals that can be used to form a likeness of the copy in another location.

Frequency, f — The number of complete cycles of a wave occurring in a unit of time.

Frequency modulation — A method of superimposing an information signal on an RF carrier wave in which the instantaneous frequency of an RF carrier wave is varied in relation to the information signal strength.

Linear or plane polarization — Describes the orientation of the electric-field component of an electromagnetic wave. The electric field can be vertical or horizontal with respect to the earth's surface, resulting in either a vertically or a horizontally polarized wave.

Magnetic field — A region through which a magnetic force will act on a magnetic object.

Magnetic force — A push or pull exerted through space by one magnetically charged object on another.

Modulation index — The ratio of the maximum frequency deviation of the modulated wave to the instantaneous frequency of the modulating signal.

Peak envelope power (PEP) — An expression used to indicate the maximum power level in a signal. It is found by squaring the RMS voltage and dividing by the load resistance.

Peak envelope voltage (PEV) — The maximum peak voltage occurring in a complex waveform.

Peak-to-peak (P-P) voltage — A measure of the voltage taken between the negative and positive peaks on a cycle.

Peak voltage — A measure of voltage on an ac waveform taken from the centerline (0 V) and the maximum positive or negative level.

Period, T — The time it takes to complete one cycle of an ac waveform.

Phase modulation — A method of superimposing an information signal on an RF carrier wave in which the phase of an RF carrier wave is varied in relation to the information signal strength.

Polarization — A property of an electromagnetic wave that describes the orientation of the electric field of the wave.

Root-mean-square (RMS) voltage — A measure of the effective value of an ac voltage.

Sawtooth wave — A waveform consisting of a linear ramp and then a return to the original value. It is made up of sine waves at a fundamental frequency and all harmonics.

Sine wave — A single-frequency waveform that can be expressed in terms of the mathematical sine function.

Single-sideband, suppressed-carrier signal — A radio signal in which only one of the two sidebands generated by amplitude modulation is transmitted. The other sideband and the RF carrier wave are removed before the signal is transmitted.

Slow-scan TV — A TV system used by amateurs to transmit pictures within a signal bandwidth allowed on the HF bands by the FCC.

Square wave — A periodic waveform that alternates between two values, and spends an equal time at each level. It is made up of sine waves at a fundamental frequency and all odd harmonics.

Thermal noise — A random noise generated by the interchange of energy within a circuit and surroundings to maintain thermal equilibrium.

Chapter 8

Signals and Emissions

There are a number of loosely related sections in this chapter. When you have studied the information in each section, use the examination questions from FCC Element 4A, listed in Chapter 10, to check your understanding of the material. If you are unable to answer a question correctly, go back and review the appropriate part of this chapter.

FCC EMISSION DESIGNATORS

The FCC uses a special system to specify the types of signals (emissions) permitted to amateurs and other users of the radio spectrum. Each emission designator has three digits; Table 8-1 shows what each character in the designator stands for. The designators begin with a letter that tells what type of modulation is being used. The second character is a number that describes the signal used to modulate the carrier, and the third character specifies the type of information being transmitted.

Some of the more common combinations are:

- NØN — Unmodulated carrier
- A1A — Morse code telegraphy using amplitude modulation
- A3E — Double-sideband, full-carrier, amplitude-modulated telephony
- J3E — Amplitude-modulated, single-sideband, suppressed-carrier telephony
- F3E — Frequency-modulated telephony
- G3E — Phase-modulated telephony
- F1B — Telegraphy using frequency-shift keying without a modulating audio tone (FSK RTTY). F1B is designed for automatic reception.
- F2B — Telegraphy produced by modulating an FM transmitter with audio tones (AFSK RTTY). F2B is also designed for automatic reception.

Facsimile and Television Emission Designators

Facsimile is the transmission of fixed images or pictures by electronic means, with the intent to reproduce the images in a permanent form. By contrast, television is the transmission of transient images of fixed or moving objects. For further discussion of facsimile and *slow-scan television* see Chapter 2.

There are several emission designators that may be used for facsimile and television signals, depending on how the signals are produced:

- A3C for full-carrier, amplitude-modulated facsimile signals with a single information channel.
- F3C for frequency-modulated facsimile signals with a single information channel.
- A3F for full-carrier, amplitude-modulated television signals with a single information channel.
- C3F for vestigial-sideband TV signals.
- J3F for single-sideband, suppressed-carrier TV signals.

Table 8-1

Partial List of WARC-79 Emissions Designators

(1) First Symbol — Modulation Type

Unmodulated carrier	N
Double sideband full carrier	A
Single sideband reduced carrier	R
Single sideband suppressed carrier	J
Vestigial sidebands	C
Frequency modulation	F
Phase modulation	G
Various forms of pulse modulation	P, K, L, M, Q, V, W, X

(2) Second Symbol — Nature of Modulating Signals

No modulating signal	0
A single channel containing quantized or digital information without the use of a modulating subcarrier	1
A single channel containing quantized or digital information with the use of a modulating subcarrier	2
A single channel containing analog information	3
Two or more channels containing quantized or digital information	7
Two or more channels containing analog information	8

(3) Third Symbol — Type of Transmitted Information

No information transmitted	N
Telegraphy — for aural reception	A
Telegraphy — for automatic reception	B
Facsimile	C
Data transmission, telemetry, telecommand	D
Telephony	E
Television	F

An examination of part 97.61 of the FCC rules reveals that these emission types go together. Where one is permitted, they are all permitted. Bandwidth restrictions for these modes are covered in part 97.65 of the rules.

When it comes to emissions designators, things are not always what they seem. *Slow-scan* TV signals consist of a series of audio tones that correspond to sync and picture-brightness levels. When SSTV is sent using an SSB transmitter, the emission is actually frequency modulation, although with the new designators, J3F is the appropriate symbol. See Chapter 2 for details. Fast-scan TV, by contrast, is normally amplitude modulation or vestigial sideband (like broadcast TV). At microwave frequencies, FM is also used for fast-scan TV in the amateur bands — as it is in the TV satellites.

If you got your Technician or General Class license before 1983 or so, you may be familiar with a different set of emission designators. Even if you got your license recently, you have probably heard other hams on the air using a set of two-letter emission designators, or seen them in an Amateur Radio book or magazine. Under this old system, for example, CW was designated A1 emission, FM telephony was F3, AM double-sideband, full-carrier telephony was A3, and amplitude-modulated television was A5. After the World Administrative Radio Conference (WARC) in 1979, the FCC began a gradual phase-in of the new emission designators, and in 1985 Part 97 was revised to express the amateur mode allocations in the new designators. These new designators allow a much more specific description of the signals and the method used to produce a given type of emission.

[Study FCC examination questions with numbers that begin 4AH-1. Review this section as needed.]

MODULATION METHODS

To pass your Advanced class exam, you will need to know what type of modulator circuit is used to produce the various types of radio signals. In Chapter 7, Practical Circuits, we described the circuit and operation of several types of modulator. If you find that you don't remember the circuit details for these modulator types, you should go back to the appropriate section in that chapter. Block diagrams of various transmitter systems are included in this section to help you understand how the pieces fit together.

Amplitude Modulation

Double-sideband, full-carrier, *amplitude modulation* (emission A3E) can be realized by simply modulating the supply voltage to an amplifier stage. See Fig. 8-1. The two signals will be mixed, so the output from the modulated stage will include the input radio frequency, the sum of the two signals (upper sideband), the difference between them (lower sideband) and the modulating frequency. One way to do this is to apply the modulating signal to the plate or collector supply voltage of a class-C RF amplifier. Other methods require the modulating signal to be applied to the grid or base circuit of a class-A or -AB amplifier. The main thing to remember is that the amplitude or strength of the output signal is changing in step with the amplitude of a modulating signal. If the frequency of that signal is also changing (as it would be for your voice), then the sideband frequencies are also changing.

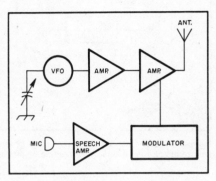

Fig. 8-1 — Block diagram of an amplitude-modulated transmitter.

Single Sideband

A *single-sideband, suppressed-carrier signal* (emission J3E) is much like an AM signal, with one important exception. By transmitting only one of the sidebands, and eliminating the carrier, SSB occupies a much smaller bandwidth. It is the most-used method for transmitting voice signals on the HF amateur bands.

SSB signals can be generated in a two-step process. In the first step a double-sideband, suppressed-carrier signal is generated in a balanced modulator. Remember that in a balanced modulator the input AF and RF signals do not appear at the output — only the sum (upper sideband) and difference (lower sideband) frequencies appear. In the second step, the unwanted sideband is filtered out, leaving only the desired one. Fig. 8-2A illustrates the essentials of such a sideband transmitter. Some transmitters use one crystal with the oscillator, but switch in a different filter to eliminate the other sideband.

There is another way to generate an SSB signal, called the phasing method. Instead of using filters to remove the unwanted sideband, the signal phase has to be adjusted carefully so the carrier, audio signal and unwanted sideband are canceled when out-of-phase components are added. This technique was popular in the early days of SSB operation, but modern filter design has made it possible to build much better filters today. Therefore, the phasing technique is seldom used. Fig. 8-2B shows the basic components of a phasing generator.

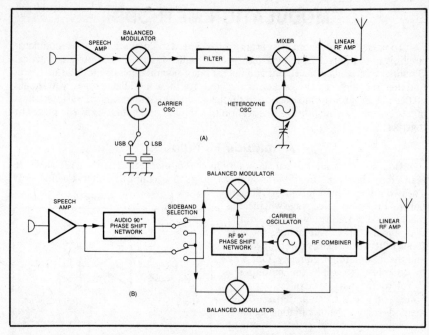

Fig. 8-2 — Part A shows the filter method of generating an SSB signal, and part B illustrates one form of the phasing method. The phasing method is not used in modern SSB equipment.

Frequency or Phase Modulation

Frequency modulation (emission F3E) operates on an entirely different principle than amplitude modulation. With FM, the signal is varied above and below the carrier frequency at a rate equal to the modulating-signal frequency. For example, if a 1000-Hz tone is used to modulate a transmitter, the carrier frequency will vary above and below the center frequency 1000 times per second. The amount of frequency change, however, depends on the instantaneous amplitude of the modulating signal. This frequency change is called *deviation*. A certain signal might produce a 5-kHz deviation. If another signal, with only half the amplitude of the first, were used to modulate the transmitter, it would produce a 2.5-kHz deviation. From this example, you can see that the deviation is proportional to the modulating signal amplitude.

Direct FM can be produced by a reactance modulator. The reactance modulator is connected to the RF tank circuit of an oscillator in such a way as to act as a variable inductance or capacitance. As a modulating signal is applied, the oscillator frequency is varied. Fig. 8-3A shows how such a system is arranged.

Phase modulation (indirect FM or emission G3E) can be realized by using a phase modulator. A phase modulator is similar to a reactance modulator in that it appears to be a variable inductance or capacitance when modulation is applied. The difference between the two is where they are found in the transmitter. Whereas the reactance modulator controls an oscillator, the phase modulator acts on a buffer or amplifier stage. This arrangement is shown at Fig. 8-3B.

FM Terminology

You will need to know two terms that refer to FM systems and operation: *devia-*

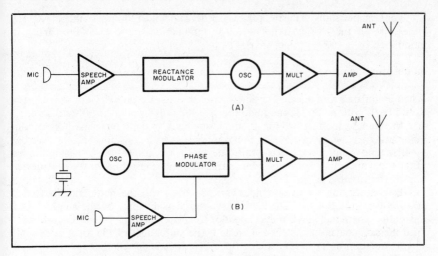

Fig. 8-3 — Direct frequency modulation (FM) is shown at A; indirect (PM) at B.

tion ratio and *modulation index*. They may seem to be almost the same — indeed, they are closely related. Pay special attention to the equations given to calculate these quantities, and you should have no problems.

Deviation Ratio

In an FM system the ratio of the maximum carrier-frequency deviation to the highest modulating frequency is called the deviation ratio. It is a constant value for a given system, calculated by:

$$\text{deviation ratio} = \frac{D_{max}}{M} \qquad \text{(Eq. 8-1)}$$

where

D_{max} = peak deviation in hertz (half the difference between the maximum and minimum carrier-frequency values at 100% modulation)

M = maximum modulating frequency in hertz

In the case of narrow-band FM, peak deviation at 100% modulation is 5 kHz. The maximum modulating frequency is 3 kHz. Therefore

$$\text{deviation ratio} = \frac{5 \text{ kHz}}{3 \text{ kHz}} = 1.67$$

Notice that since both frequencies were given in kilohertz we did not have to change them to hertz before doing the calculation. The important thing is that they both be in the same units.

Modulation Index

The ratio of the maximum carrier-frequency deviation to the modulating frequency is called the modulation index. That is

$$\text{modulation index} = \frac{D_{max}}{m} \qquad \text{(Eq. 8-2)}$$

where

D_{max} = peak deviation in hertz

m = modulating frequency in hertz

Signals and Emissions   **8-5**

For example, suppose the peak frequency deviation of an FM transmitter is 3000 Hz either side of the carrier frequency. The modulation index when the carrier is modulated by a 1000-Hz sine wave is

$$\text{modulation index} = \frac{3000 \text{ Hz}}{1000 \text{ Hz}} = 3$$

When modulated with a 3000-Hz sine wave with the same peak deviation, the index would be 1; with a 100-Hz modulating wave the index would be 30, and so on.

In a phase modulator, the modulation index is constant regardless of the modulating frequency, as long as the amplitude is held constant. In other words, a 2-kHz tone will produce twice as much deviation as a 1-kHz tone if the amplitudes of the tones are equal. This may seem confusing at first; just think of the peak deviation as the variable in Eq. 8-2.

By contrast, the modulation index varies inversely with the modulating frequency in a frequency modulator. The actual deviation depends only on the amplitude of the modulating signal and is independent of frequency. Thus, a 2-kHz tone will produce the same deviation as a 1-kHz tone if the amplitudes of the tones are equal. The modulation index in the case of the 1-kHz tone is double that for the case of the 2-kHz tone.

[At this point you should turn to Chapter 10 and study FCC examination questions with numbers between 4AH-2.1 and 4AH4.5. Review this section as needed.]

ELECTROMAGNETIC WAVES

All *electromagnetic waves* are moving fields of *electric* and *magnetic* force. Their lines of force are at right angles to each other, and are also both perpendicular to the direction of travel. See Fig. 8-4. They can have any position with respect to the earth. The plane containing the continuous lines of electric and magnetic force is called the wave front. Another way of visualizing this concept is to think of the wave front as being a fixed point on the moving wave.

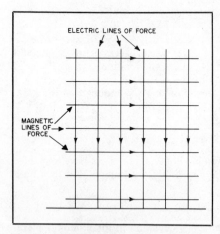

Fig. 8-4 — Representation of electric and magnetic lines of force in a radio wave. Arrows indicate instantaneous directions of the fields for a wave traveling toward you, out of the page. Reversing the direction of one set of lines would reverse the direction of travel, but if you reversed the direction of both sets the wave would still be coming out of the page.

Electromagnetic Radiation

Electricity requires a conductor to carry an electron current through a circuit. Electromagnetic waves move easily through the vacuum of free space. The *electric* and *magnetic fields* that constitute the wave do not require a conductor to carry them.

The medium in which electromagnetic waves travel has a marked influence on their speed of movement. In empty space electromagnetic waves travel at the same speed as light, 300,000,000 meters per second or about 186,000 miles per second. It is slightly less in air, and it varies somewhat with temperature and humidity, depending on the frequency. It is much less in other *dielectric materials* (insulators), where the speed is inversely proportional to the square root of the *dielectric constant* of the material.

Radio waves travel through dielectric materials with ease. Waves cannot penetrate a good conductor, however. Instead of penetrating the conductor as they encounter it, the magnetic field generates current in the conductor surface. These induced currents are called eddy currents.

Wave Polarization

Polarization refers to the direction of the electric lines of force of a radio wave. See Fig. 8-5. If the electric lines of force are parallel to the earth, we call this a horizontally polarized radio wave. In a horizontally polarized wave, the electric lines of force are horizontal and the magnetic lines are vertical. A radio wave is vertically polarized if its electric lines of force are perpendicular to the earth (vertical). In this case the magnetic lines are horizontal.

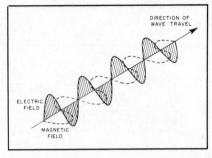

Fig. 8-5 — Representation of the magnetic and electric fields of a vertically polarized radio wave. In this diagram, the electric field is in a vertical plane and the magnetic field is in a horizontal plane.

For the most part, polarization is determined by the type of transmitting antenna used, and its orientation. On one hand, for example, a Yagi antenna with its elements parallel to the earth's surface transmits a horizontally polarized radio wave. On the other hand, an amateur mobile whip antenna, mounted vertically on an automobile, radiates a vertically polarized wave.

It is possible to generate waves with rotating field lines. This condition, where the electric field lines are continuously rotating through horizontal and vertical orientations, is called *circular polarization*. It is particularly helpful to use circular polarization in satellite communication, where polarization tends to shift.

Polarization that does not rotate is called *linear* or *plane polarization*. Horizontal and vertical polarization are examples of linear polarization. (In space, of course, horizontal and vertical have no convenient reference.) Circular polarization is usable with linearly polarized antennas at the other end of the circuit. There will be some small loss in this case, however. If you use a vertically polarized antenna to receive a horizontally polarized radio wave (or vice versa) over a line-of-sight ground-wave path, you can expect the received signal strength to be reduced by more than 20 dB as compared to using an antenna with the same polarization as the wave. With propagation paths that use sky waves, this effect may disappear completely.

[Now study those FCC examination questions with numbers between 4AH-5.1 and 4AH-6.7 Review this section as needed.]

AC WAVEFORMS

Sine Waves

The basic ac waveform is the *sine wave*. A sine wave represents a single *frequency*. To visualize a sine wave, let's imagine a wheel, like a bicycle wheel. We will paint a dot on the edge of the wheel at one point, so we can watch that spot as the wheel spins. If you look at the wheel edge, the spot will just seem to move up and down, as illustrated in Fig. 8-6A. B pictures the wheel from one side, with the spot shown stopped at several points around the circle. If you can imagine the pattern at A as the wheel moves sideways to the right, the spot will trace a sine wave, as shown at C. This sine wave also represents the output from an ac generator, or alternator as

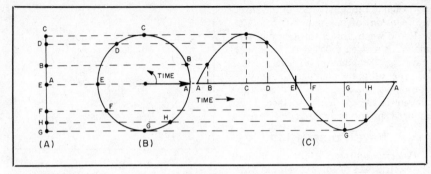

Fig. 8-6 — This diagram illustrates the relationship between a sine wave and an object rotating in a circle. You can see how various points on the circle relate to sine-wave values.

it is called. One full rotation corresponds to a complete cycle (360 degrees).

Fig. 8-6 is not only a mechanical analogy of alternator operation; it is a mathematical model as well. In the mathematical model, a line drawn from the axle to our paint spot is a rotating vector. The graph describes the vector ordinate (value along the Y or vertical axis) as it varies with time. Twice during each cycle a sine wave passes through zero; once while going positive, and once while going negative.

The time required to complete one cycle is called the *period, T*. The frequency of the sine wave is the reciprocal of the period:

$$f = \frac{1}{T} \tag{Eq. 8-3}$$

Sawtooth Waves

A *sawtooth wave*, as shown in Fig. 8-7, is so named because it closely resembles the teeth on a saw blade. It is characterized by a rise time significantly faster than the fall time (or vice versa). A sawtooth wave is made up of a sine wave at the fundamental frequency and sine waves at all the harmonic frequencies as well. When a sawtooth voltage is applied to the horizontal deflection plates of an oscilloscope, the electron beam sweeps slowly across the screen during the slowly changing portion of the waveform and then flies quickly back during the rapidly changing portion of the signal. This type of waveform is desired to obtain a linear sweep in an oscilloscope.

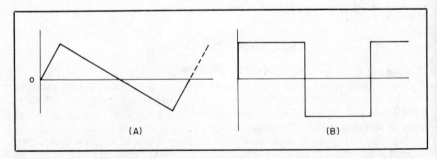

Fig. 8-7 — Any waveform that is not a pure sine wave contains harmonics. Sawtooth waves (A) consist of both odd and even harmonics as well as the fundamental. Square waves (B) consist of only fundamental and odd-harmonic frequencies.

Square Waves

A *square wave* is one that abruptly changes back and forth between two voltage levels and remains an equal time at each level. See Fig. 8-7B. (If the wave spends an unequal time at each level, it is known as a rectangular wave.) A square wave is made up of sine waves at the fundamental and all the odd harmonic frequencies.

[Study FCC examination questions with numbers that begin 4AH-7. Review this section as needed.]

AC MEASUREMENTS

The time dependence of alternating-current waveforms raises questions about defining and measuring values of voltage, current and power. Because these parameters change from one instant to the next, one might wonder, for example, which point on the cycle characterizes the voltage or current for the entire cycle. Since the wave is positive for exactly the same time it is negative in value, you might even wonder if the value shouldn't be zero. Actually, the average dc voltage and current are zero! A dc meter connected to an ac voltage would read zero, although you may be able to notice some slight flutter on a sensitive meter. To get some idea of how useful the ac voltage might be, we will have to connect a diode in series with the meter lead and use a specially calibrated scale.

Ac Voltage and Current

When viewing a sine wave on an oscilloscope, the easiest dimension to measure is the total vertical displacement, or *peak-to-peak* (P-P) *voltage*. The maximum positive or negative potential is called the *peak voltage*. In a symmetrical waveform it has half the value of the peak-to-peak amplitude.

When an ac voltage is applied to a resistor, the resistor will dissipate energy in the form of heat, just as if the voltage were dc. The dc voltage that would cause identical heating in the ac-excited resistor is called the *root-mean-square* (RMS) or effective value of the ac voltage. The phrase root-mean-square describes the mathematical process of actually calculating the effective value. The method involves squaring the peak value for a large number of points along the waveform (a calculus procedure), then finding the average of the squared values, and taking the square root of that number. The RMS voltage of any waveform can also be determined by measuring the heating effect in a resistor. For sine waves, the following relationships hold:

$$V_{peak} = V_{RMS} \times \sqrt{2} = V_{RMS} \times 1.414 \qquad \text{(Eq. 8-4)}$$

and

$$V_{RMS} = \frac{V_{peak}}{\sqrt{2}} = V_{peak} \times 0.707 \qquad \text{(Eq. 8-5)}$$

If we consider only the positive or negative half of a cycle, it is possible to calculate an average value for a sine-wave voltage. Meter movements respond to this average value rather than either the peak or RMS values of a voltage because of the inertia inherent in the needle and magnet. Often, the meter has a scale that is calibrated to read RMS values, even though the needle is actually responding to the average value. That is okay, as long as the waveform you are measuring is a pure sine wave. Other, more complex, waveforms will not give a true reading, however. The mathe-

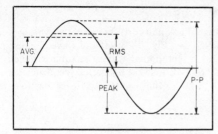

Fig. 8-8 — An illustration showing ac voltage and current measurement terms for a pure sine wave.

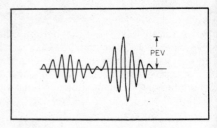

Fig. 8-9 — A complex waveform, made up of several individual sine-wave signals. Peak envelope voltage (PEV) is an important parameter for a composite waveform.

matical relationships between average, peak and RMS values for a sine wave are given by:

$$V_{avg} = V_{peak} \times 0.636 \qquad \text{(Eq. 8-6)}$$

and

$$V_{avg} = V_{RMS} \times 0.899 \qquad \text{(Eq. 8-7)}$$

Unless otherwise specified or obvious from the context, ac voltage is rendered as an RMS value. For example, the household 117-V ac outlet provides 117-V RMS, 165.5-V peak and 331-V P-P. The voltage at your household outlets varies with the amount of load that the power company must supply. It will vary around the nominal 117-V value, and is sometimes even specified as 110. Of course this means that the peak and P-P values also vary. Fig. 8-8 illustrates the voltage parameters of a sine wave.

The significant dimension of a multitone signal (a complex waveform) is the *peak envelope voltage* (PEV), shown in Fig. 8-9. PEV is important in calculating the power in a modulated signal, such as that from an amateur SSB transmitter.

All that has been said about voltage measurements applies also to current (provided the load is resistive) because the waveshapes are identical.

Ac Power

The terms RMS, average and peak have different meanings when they refer to ac power. The reason is that while voltage and current are sinusoidal functions of time, power is the product of voltage and current, and this product is a sine squared function. The calculus operations that define RMS, average and peak values will naturally yield different results when applied to this new function. The relationships between ac voltage, current and power are as follows:

$$V_{RMS} \times I_{RMS} = P_{avg} \qquad \text{(Eq. 8-8)}$$

Note that this calculation does not give a value for RMS power. The average power used to heat a resistor is equal to the dc power required to produce the same heat. RMS power has no physical significance!

For continuous sine wave signals:

$$V_{peak} \times I_{peak} = P_{peak} = 2 \times P_{avg} \qquad \text{(Eq. 8-9)}$$

Unfortunately, the situation is more complicated in radio work. We seldom have a steady sine-wave signal being produced by a transmitter. The waveform varies with time, in order to carry some useful information for us. The peak power output of a radio transmitter, then, is the power averaged over the RF cycle having the greatest amplitude. Modulated signals are not purely sinusoidal because they are composites

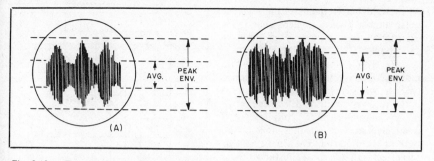

Fig. 8-10 — Two envelope patterns that show the difference between average and peak levels. In each case, the RF amplitude (current or voltage) is plotted as a function of time. In B, the average level has been increased. That will raise the average output power compared to the peak value.

of two or more audio tones. However, the cycle-to-cycle variation is small enough that sine-wave measurement techniques produce accurate results. In the context of radio signals, then, peak power means maximum average power. *Peak envelope power* (PEP) is the parameter most often used to express the maximum signal level. To compute the PEP of a waveform such as that sketched in Fig. 8-9, multiply the PEV by 0.707 to obtain the RMS value, square the result and divide by the load resistance.

SSB Power

Envelope peaks occur only sporadically during voice transmission and have no relationship with meter readings. The meters respond to the amplitude (current or voltage) of the signal averaged over several cycles of the modulation envelope.

The ratio of peak-to-average amplitude varies widely with voices of different characteristics. In the case shown in Fig. 8-10, the average amplitude (found graphically) is such that the peak-to-average ratio of amplitudes is almost 3:1. Typical ratio values range from 2:1 to more than 10:1. So the PEP of an SSB signal may be about 2 or 3 times greater than the average power output. It may even be more than that, depending on the voice characteristics of the person speaking into the microphone.

[Turn to Chapter 10 and study FCC examination questions with numbers that begin 4AH-8 and 4AH-9. Review this section as needed.]

SIGNAL-TO-NOISE RATIO

Receiver noise performance is established primarily in the RF amplifier and/or mixer stages. Low-noise, active devices should be used in the receiver front end to obtain good performance. The unwanted noise, in effect, masks the weaker signals and makes them difficult or impossible to copy. Noise generated in the receiver front end is amplified in the succeeding stages along with the signal energy. Therefore, in the interest of sensitivity, internal noise should be kept as low as possible.

Don't confuse external noise (man-made and atmospheric noise, which comes in on the antenna lead) with receiver noise during discussions of noise performance. The ratio of external noise to the incoming signal level does have a lot to do with reception. It is because external noise levels are quite high on the 160 through 20-meter bands that emphasis is seldom placed on low-internal-noise receivers for those bands. As the operating frequency is increased from 15 meters up through the microwave spectrum, however, the matter of receiver noise becomes a primary consideration.

At these higher frequencies the receiver noise almost always exceeds that from external sources, especially at 2 meters and above.

Receiver noise is produced by the movement of electrons in any substance (such as wires, resistors and transistors) in the receiver circuitry. Electrons move in a random fashion colliding with relatively immobile ions that make up the bulk of the material. The final result of this effect is that in most substances there is no net current in any particular direction on a long-term average, but rather a series of random pulses. These pulses produce what is called thermal-agitation noise, or simply *thermal noise*.

Thermal-noise power is directly proportional to bandwidth and absolute temperature (in kelvins). For that reason, narrow-band systems exhibit better noise performance than do wide-band systems. For extremely low-noise operation some amplifiers are cooled with liquid air or nitrogen. That practice is hazardous, however, and is not used by amateurs. Noise temperature, noise factor and noise figure are all measures of this thermal noise. Fig. 8-11 shows the relationship between noise temperature in kelvins and noise figure in decibels.

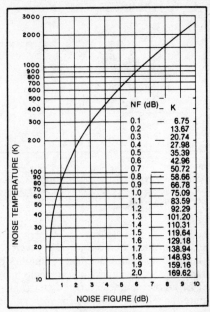

Fig. 8-11 — Relationship between noise figure and noise temperature.

[Now study FCC examination questions with numbers that begin 4AH-10. Review this section as needed.]

Key Words

Antenna — An electric circuit designed specifically to radiate the energy applied to it in the form of electromagnetic waves. An antenna is reciprocal; a wave moving past it will induce a current in the circuit also. Antennas are used to transmit and receive radio waves.

Antenna bandwidth — A range of frequencies over which an antenna will perform well. Antenna bandwidth is usually specified as a range of frequencies where the antenna SWR will be below some given value.

Antenna efficiency — The ratio of the radiation resistance to the total resistance of an antenna system.

Base loading — The technique of inserting a coil with specific reactance at the bottom of a vertical antenna in order to cancel the capacitive reactance of the antenna, producing a resonant antenna system.

Beamwidth — As related to directive antennas, the width (measured in degrees) of the major lobe between the two directions at which the relative power is one half (– 3 dB) its value at the peak of the lobe.

Center loading — A technique for adding a series inductor at or near the center of an antenna element in order to cancel the capacitive reactance of the antenna. This technique is usually used with elements that are less than ¼ wavelength.

Dielectric — An insulating material. A dielectric is a medium in which it is possible to maintain an electric field with little or no additional energy supplied after the field has been established.

Dielectric constant — Relative figure of merit for an insulating material. This is the property that determines how much electric energy can be stored in a unit volume of the material per volt of applied potential.

Dipole — An antenna with two elements in a straight line that are fed in the center; literally, two poles. For amateur work, dipoles are usually operated at half-wave resonance.

Director — A parasitic element located in front of the driven element of a beam antenna. It is intended to increase the strength of the signals radiated from the front of the antenna. Typically about 5% shorter than the driven element.

Driven element — Any antenna element connected directly to the feed line.

Folded dipole — An antenna consisting of two (or more) parallel, closely spaced half-wave wires connected at their ends. One of the wires is fed at its center.

Frequency — A property of an electromagnetic wave that refers to the number of complete alternations (or oscillations) made in one second.

Gain — An increase in the effective power radiated by an antenna in a certain desired direction. This is at the expense of power radiated in other directions.

Loading coil — An inductor that is inserted in an antenna element or transmission line for the purpose of producing a resonant system at a specific frequency.

Major lobe of radiation — A three-dimensional area in the space around an antenna that contains the maximum radiation peak. The field strength decreases from the peak level, until a point is reached where it starts to increase again. This area is known as the major lobe.

Minor lobe of radiation — Those areas of an antenna pattern where there is some increase in radiation, but not as much as in the major lobe. Minor lobes normally appear at the back and sides of the antenna.

Parasitic element — An antenna element not directly connected to the feed line, but which affects the radiation pattern and feed-point impedance of the antenna.

Radiation resistance — The equivalent resistance that would dissipate the same amount of power as is radiated from an antenna. It is calculated by dividing the radiated power by the square of the RMS antenna current.

Reflector — A parasitic antenna element that is located behind the driven element to enhance forward directivity. The reflector is usually about 5% longer than the driven element.

Self-resonant antenna — An antenna that is to be used on the frequency at which it is resonant, without the use of any added inductive or capacitive loading elements.

Top loading — The addition of inductive reactance (a coil) or capacitive reactance (a capacitance hat) at the end of a driven element opposite the feed point. It is intended to increase the electrical length of the radiator.

Traps — Parallel LC networks inserted in an antenna element to provide multiband operation.

Velocity factor — An expression of how fast a radio wave will travel through a material. It is usually stated as a fraction of the speed the wave would have in free space (where the wave would have its maximum velocity). Velocity factor is also sometimes specified as a percentage of the speed of a radio wave in free space.

Wavelength — The distance between two points with corresponding phase on two consecutive cycles of a wave.

Chapter 9

Antennas and Feed Lines

T here are several sections in this chapter. These sections cover various antenna and feed-line topics. When you have studied the information in each section, use the examination questions to check your understanding of the material. If you are unable to answer a question correctly, go back and review the appropriate part of this chapter.

VELOCITY FACTOR AND ELECTRICAL LENGTH OF A TRANSMISSION LINE

Radio waves travel through space at a speed of 300,000,000 meters per second (approximately 186,000 miles per second). If a wave travels through anything but a vacuum, its speed is always less than that.

Waves used in radio communication may have frequencies from about 10,000 to several billion hertz (Hz). Suppose the *frequency* of the wave shown in Fig. 9-1 is 30,000,000 Hz, or 30 megahertz (MHz). One cycle is completed in 1/30,000,000 second (30 µs). This time is called the period of the wave. The wave is traveling at 300,000,000 meters per second, so it will move only 10 meters during the time the current is going through one complete cycle. The electromagnetic field 10 meters away from the source is caused by the current that was flowing one period earlier in time. The field 20 meters away is caused by the current that was flowing two periods earlier, and so on.

Wavelength is the distance between two points of the same phase (for example, peaks) in consecutive cycles. This distance must be measured along the direction of wave travel. In the example found in the previous paragraph, the wavelength is 10 meters. The formula for wavelength is:

$$\lambda = \frac{v}{f} \qquad \text{(Eq. 9-1)}$$

where
  λ = wavelength
  v = velocity of wave
  f = frequency of wave

Equations normally specify basic units, such as velocity in m/s or ft/s and fre-

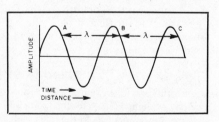

Fig. 9-1 — The instantaneous amplitude of both fields (electric and magnetic) varies sinusoidally with time, as shown in this graph. Since the fields travel at constant velocity, the graph also represents the in-stantaneous distribution of field intensity along the wave path. The distance between two points of equal phase, such as A-B and B-C, is the length of the wave.

quency in Hz. When we are working with very large or very small numbers it is sometimes more convenient to learn a formula that includes a simplified number and specifies units in more common multiples, such as frequency in MHz. For waves traveling in free space, the formula to calculate wavelength is:

$$\lambda(\text{meters}) = \frac{300}{f\ (\text{MHz})} \qquad \text{(Eq. 9-2)}$$

or

$$\lambda(\text{feet}) = \frac{984}{f\ (\text{MHz})} \qquad \text{(Eq. 9-3)}$$

Wavelength in a Wire

An alternating voltage applied to a line would give rise to the sort of current flow shown in Fig. 9-2. If the frequency of the ac voltage is 10 MHz, each cycle will occupy 0.1 microsecond. Therefore, a complete current cycle will be present along each 30 meters of line (assuming free-space velocity). This distance is one wavelength. Current observed at B occurs just one cycle later in time than the current at A. To put it another way, the current initiated at A does not appear at B, one wavelength away, until the applied voltage has had time to go through a complete cycle.

In Fig. 9-2, the series of drawings shows how the instantaneous current might appear if we could take snapshots of it at quarter-cycle intervals. The current travels out from the input end of the line in waves.

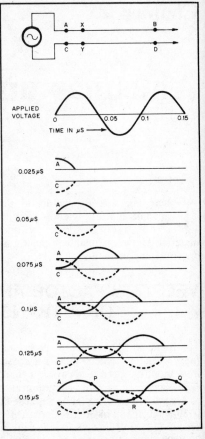

Fig. 9-2 — Instantaneous current along a transmission line at successive time intervals. The period of the wave (the time for one cycle) is 0.1 microsecond.

At any selected point on the line, the current goes through its complete range of ac values in the time of one cycle just as it does at the input end. Therefore an ammeter inserted in the line would read the same current at any point along the line (assuming no losses).

Velocity of Propagation

In the previous example, we assumed that energy traveled along the line at the velocity of light. The actual velocity is very close to that of light if the insulation between the conductors of the line is solely air. The presence of *dielectrics* other than air reduces the velocity, since electromagnetic waves travel more slowly in materials other than a vacuum. Because of this, the length of line in one wavelength will depend on the velocity of the wave as it moves along the line.

The ratio of the actual velocity at which a signal travels along a line to the speed of light in a vacuum is called the *velocity factor*. For example, the velocity factor of several types of coaxial cable is 0.66. The velocity factor is related to the *dielectric constant*, ϵ, by:

Table 9-1

Characteristics of Commonly Used Transmission Lines

Type of line	Z_0 Ohms	Vel %	pF per foot	OD (inches)	Diel. Material	Max. Operating Volts (RMS)
RG-8/U	52.0	66	29.5	0.405	PE	4000
RG-8/U Foam	50.0	80	25.4	0.405	Foam PE	1500
RG-58/U	53.5	66	28.5	0.195	PE	1900
RG-58/U Foam	53.5	79	28.5	0.195	Foam PE	600
RG-59/U	73.0	66	21.0	0.242	PE	2300
RG-59/U Foam	75.0	79	16.9	0.242	Foam PE	800
RG-141/U	50.0	70	29.4	0.190	PTFE	1900
RG-174/U	50.0	66	30.8	0.1	PE	1500
Aluminum Jacket Foam Dielectric						
1/2 inch	50.0	81	25.0	0.5		2500
3/4 inch	50.0	81	25.0	0.75		4000
7/8 inch	50.0	81	25.0	0.875		4500
1/2 inch	75.0	81	16.7	0.5		2500
3/4 inch	75.0	81	16.7	0.75		3500
7/8 inch	75.0	81	16.7	0.875		4000
Open wire	—	97	—	—		—
75-ohm trans- mitting twin lead	75.0	67	19.0	—		—
300-ohm twin lead	300.0	80	5.8	—		—
300-ohm tubular	300.0	77	4.6	—		—
Open wire, TV type						
1/2 inch	300.0	95	—	—		—
1 inch	450.0	95	—	—		—

Dielectric Designation	Name	Temperature Limits
PE	Polyethylene	$-65°$ to $+80°C$
Foam PE	Foamed Polyethylene	$-65°$ to $+80°C$
PTFE	Polytetrafluoroethylene (Teflon)	$-65°$ to $+250°C$

$$V = \frac{1}{\sqrt{\epsilon}}$$ (Eq. 9-4)

where

V = velocity factor
ϵ = dielectric constant

Electrical Length

The electrical length of a transmission line (or antenna) is not the same as its physical length. The electrical length is measured in wavelengths at a given frequency. To calculate the physical length of a transmission line that is electrically one wavelength, use the formulas

$$\text{Length (m)} = \frac{300\ V}{f\ (\text{MHz})}$$ (Eq. 9-5)

or

$$\text{Length (ft)} = \frac{984\ V}{f\ (\text{MHz})}$$ (Eq. 9-6)

where

f = operating frequency (in MHz)
V = velocity factor

Suppose you want a section of RG-8/U coaxial cable that is one-half wavelength at 3.8 MHz. What is its physical length? The answer depends on the dielectric used in the coaxial cable. RG-8/U is manufactured with polyethylene or foamed polyethylene dielectric; velocity factors for the two versions are 0.66 and 0.80, respectively. The length in feet is then:

$$\text{Length (ft)} = \frac{0.5 \times 984 \times 0.66}{3.8} = 85.5 \text{ ft (polyethylene)}$$

or

$$\text{Length (ft)} = \frac{0.5 \times 984 \times 0.80}{3.8} = 103.6 \text{ ft (foamed polyethylene)}$$

Table 9-1 lists velocity factors for some other common feed lines.

In review, the lower the velocity factor, the slower a radio-frequency wave moves through the line. The lower the velocity factor, the shorter a line is for the same electrical length at a given frequency. One wavelength in a practical line is always shorter than a wavelength in free space.

[Before proceeding, study FCC examination questions with numbers that begin 4AI-8 and 4AI-9. Review this section as needed.]

DRIVEN ELEMENTS

A *driven element* is an antenna element that is supplied power directly from the transmitter through a wire or conductor. The fundamental amateur antenna is a wire whose total electrical length is half the transmitted-signal wavelength. This antenna is so basic, in fact, that it is the unit from which many more complex antennas are constructed. Amateurs most often call this antenna the half-wave dipole antenna.

The physical length of a half wavelength in free space is:

$$\text{Length (feet)} = \frac{492}{f \text{ (MHz)}} \qquad \text{(Eq. 9-7)}$$

The resonant length of a real-life half-wave antenna will not be exactly equal to the half wave in space, but depends on the thickness of the conductor in relation to the length. The length of an actual half-wavelength wire antenna is approximately:

$$\text{Length (feet)} = \frac{468}{f \text{ (MHz)}} \qquad \text{(Eq. 9-8)}$$

[Now study FCC examination questions with numbers that begin 4AI-5. Review the material in this section as needed.]

CURRENT AND VOLTAGE DISTRIBUTION

If the wire in an antenna were infinitely long, the charge (voltage) and the current (an electric current is simply a charge in motion) would both slowly decrease in amplitude with distance from the source. The slow decrease would result from dissipation of energy in the form of radio waves and in heating the wire because of its resistance. If the wire is short, the charge is reflected when it reaches the far end, however. When radio-frequency energy excites a half-wave antenna, there is, of course, not just a single charge but a continuous supply of energy, varying in voltage according to a sine-wave cycle. We might consider this as a series of charges, each of slightly different amplitude than the preceding one. When a charge reaches the end of the antenna and is reflected, the direction of current flow reverses, since the charge is now traveling in the opposite direction. The next charge is just reaching the end of the antenna, however, so we have two currents of practically the same amplitude flowing in opposite directions. The resultant current at the end of the antenna therefore is zero. As we move farther back from the end of the antenna the magnitudes of the outgoing and returning currents are no longer the same because the charges causing them have been supplied to the antenna at different parts of the RF cycle. There is less cancellation, therefore, and a measurable current exists.

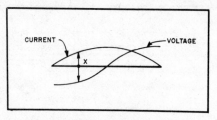

Fig. 9-3 — Current and voltage distribution on a half-wave wire. The wire is represented by the heavy line. In this conventional representation, the distance at any point (X, for instance) from the wire to the curve gives the relative current or voltage intensity at that point. The relative direction of current flow (or voltage polarity) is indicated by drawing the curve either above or below the antenna line. The voltage curve here, for example, indicates that the instantaneous polarity in one half of the antenna is opposite to that in the other half.

The greatest difference — that is, the largest resultant current — will be found to exist a quarter wavelength away from the end of the antenna. As we move back still farther from this point the current will decrease until, a half wavelength away from the end of the antenna, it will reach zero again. Thus, in a half-wave antenna the current is zero at the ends and maximum at the center.

This resultant current distribution along a half-wave wire is shown in Fig. 9-3. The distance measured vertically from the antenna wire to the curve marked "current," at any point along the wire, represents the relative amplitude of the current as measured by an ammeter at that point. This is called a standing wave of current. The instantaneous value of current at any point varies sinusoidally at the applied frequency, but its amplitude is different at various points along the wire, as shown by the curve. The standing-wave curve itself has the shape of a half sine wave, at least to a good approximation.

The voltage along the wire will behave differently; it is obviously greatest at the end since at this point we have two practically equal charges adding. As we move back along the wire, however, the outgoing and returning charges are not equal and their sum is smaller. At the quarter-wave point the returning charge is of equal magnitude but of opposite sign to the outgoing charge, since at this time the polarity of the voltage wave from the source has reversed (one-half cycle). The two voltages therefore cancel each other and the resultant voltage is zero. Beyond the quarter-wave point, away from the end of the wire, the voltage again increases, but this time with the opposite polarity.

You can observe, therefore, that the voltage is maximum at every point where the current is minimum, and vice versa. The polarity of the current or voltage reverses every half wavelength along the wire, but the reversals do not occur at the same points for both current and voltage; the respective reversals occur, in fact, at points a quarter wavelength apart. A maximum point on a standing wave is called a loop (or antinode); a minimum point is called a node.

[Before proceeding to the next section, study FCC examination questions with numbers that begin 4AI-10. Review this section as needed.]

TRAP ANTENNAS

By using tuned circuits of appropriate design strategically placed in a dipole, the antenna can be made to show what is essentially fundamental resonance at a number of different frequencies. The general principle is illustrated by Fig. 9-4. The two inner lengths of wire, X, together form a simple dipole resonant at the highest

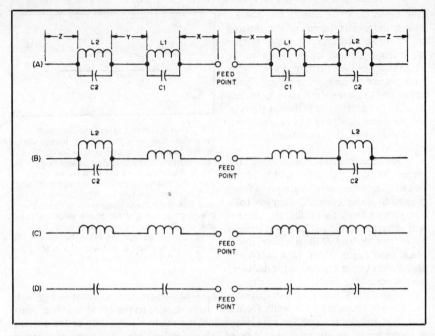

Fig. 9-4 — Development of the trap dipole for operation on fundamental-type resonance in several bands. Part A shows the basic construction of an antenna that uses two sets of traps, for operation on three frequency bands. Parts B and C show the inductive loading of the antenna on successively lower bands. D shows the capacitive loading that would result if the antenna were operated on a higher frequency, although the antenna will not normally be used on frequencies higher than the inner-dipole-section resonance.

band desired, say 14 MHz. The tuned circuits L1-C1 (called *traps*) are also resonant at this frequency, and when connected as shown offer a very high impedance to RF current of that frequency which may be flowing in the section X-X. Effectively, therefore, these two tuned circuits act as insulators for the inner dipole, and the outer sections beyond L1-C1 are inactive.

On the next lower frequency band of interest, say 7 MHz, L1-C1 shows an inductive reactance and is the electrical equivalent of a coil. If the two sections marked Y are now added and their length adjusted so that, together with the *loading coils* represented by the inductive reactance of L1-C1, the system is resonant at 7 MHz out to the ends of the Y sections. This part of the antenna is equivalent to a loaded dipole on 7 MHz and will exhibit about the same impedance at the feed point as a simple dipole for that band. The tuned circuit L2-C2 is resonant at 7 MHz and acts as a high impedance for this frequency, so the 7-MHz dipole is in turn insulated, for all practical purposes, from the remaining outer parts of the antenna.

Carrying the same reasoning one step further, L2-C2 shows inductive reactance on the next lower frequency band, 3.5 MHz, and is equivalent to a coil on that band. The length of the added sections, Z-Z, is adjusted so that, together with the two sets of equivalent loading coils indicated in part C, the whole system is resonant as a loaded dipole on 3.5 MHz. A single transmission line having a characteristic impedance of the same order as the feed-point impedance of a simple dipole can be connected at the center of the antenna. This line will be satisfactorily matched on all three bands, and so will operate at a low SWR on all three. A line of 50-ohm impedance will work just fine.

Since the tuned circuits have some inherent losses, the efficiency of this system depends on the Q of the tuned circuits. Low-loss (high-Q) coils should be used, and the capacitor losses likewise should be kept as low as possible. With tuned circuits that are good in this respect — comparable with the low-loss components used in transmitter tank circuits, for example — the reduction in efficiency as compared with the efficiency of a simple dipole is small, but tuned circuits of low Q can lose an appreciable portion of the power supplied to the antenna.

The lengths of the added antenna sections Y and Z must, in general, be determined experimentally. The length required for resonance in a given band depends on the length/diameter ratio of the antenna conductor and on the LC ratio of the trap acting as a loading coil. The effective reactance of an LC circuit at half the frequency to which it is resonant is equal to $\frac{2}{3}$ the reactance of the inductance at the resonant frequency. For example, if L1-C1 resonates at 14 MHz and L1 has an inductive reactance of 300 ohms at that frequency, the inductive reactance of the circuit at 7 MHz will be equal to $\frac{2}{3} \times 300 = 200$ ohms. The added antenna section, Y, would have to be cut to the proper length to resonate at 7 MHz with this amount of loading. Since any reasonable LC ratio can be used in the trap without affecting its performance materially at the resonant frequency, the LC ratio can be varied to control the added antenna length required. The added section will be shorter with high-L trap circuits and longer with high-C traps.

Trap dipoles have two major disadvantages. Because the trap dipole is a multi-band antenna, it can do a good job of radiating harmonics. Further, during operation on the lower frequency bands, the series inductance (loading) from the traps raises the Q of the antenna. That means less bandwidth for a given SWR limit.

[Study FCC examination questions with numbers that begin 4AI-2. Review this section as needed.]

FOLDED DIPOLE ANTENNAS

In Fig. 9-5, suppose for the moment that the upper conductor between points B and C is disconnected and removed. The system is then a simple center-fed dipole, and the direction of current flow along the antenna and line at a given instant is as given by the arrows. Then if the upper conductor between B and C is restored, the current in it will flow away from B and toward C.

This may seem confusing and be opposite to the direction you would expect the current to flow on that portion of the line. Just remember that for a sine wave, the current direction is reversed in alternate half-wave sections along a wire. Because of the way the second wire is "folded," however, the currents in the two conductors of the antenna are actually flowing in the same direction. Although the antenna physically resembles a transmission line, it is not actually a line. The antenna element merely consists of two parallel conductors carrying current in the same direction. If it were acting like a transmission line, the currents would be flowing in opposite directions. The connections at the ends of the two conductors are assumed to be of negligible length.

A half-wave dipole formed in this way will have the same directional properties and total radiation resistance as an ordinary dipole. The transmission line is connected to only one of the conductors, however. You should expect that the antenna will "look" different, with

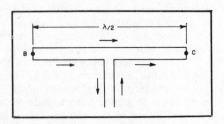

Fig. 9-5 — Direction of current flow in a folded dipole.

respect to its input impedance, as viewed by the line.

The effect on the impedance at the antenna input terminals can be visualized quite readily. The center impedance of the dipole as a whole is the same as the impedance of a single-conductor dipole — that is, approximately 73 ohms. A given amount of power will therefore cause a definite value of current, I. In the ordinary half-wave dipole this current flows at the junction of the line and antenna. In the *folded dipole* the same total current also flows, but is equally divided between two conductors in parallel. The current in each conductor is therefore I/2. Consequently, the line "sees" a higher impedance because it is delivering the same power at only half the current.

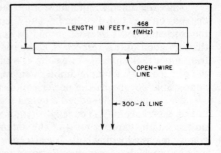

Fig. 9-6 — Construction information for a folded half-wave dipole.

Ohm's law reveals that the new value of impedance is equal to four times the impedance of a simple dipole. If more wires are added in parallel the current continues to divide between them and the terminal impedance is raised still more. This explanation is a simplified one based on the assumption that the conductors are close together and have the same diameter.

Another advantage of the folded dipole is that it has a low SWR over a wider frequency range than a normal dipole. This increased bandwidth can be accounted for if you understand that as the operating frequency varies from the resonant frequency, part of the dipole element begins to act as a shorted feed line. The feed line reactance is of the type opposite to that of the antenna elements, so the reactances tend to cancel.

The two-wire system shown in Fig. 9-6 is an especially useful one because the input impedance is so close to 300 ohms that it can be fed directly with 300-ohm twin-lead or open-wire line without any other matching arrangement.

[Study FCC examination questions with numbers that begin 4AI-7. Review this section as needed.]

RADIATION RESISTANCE

The energy supplied to an antenna is dissipated in the form of radio waves and in heat losses in the wire and nearby dielectrics. The radiated energy is the useful part, and as far as the antenna is concerned it represents a loss just as much as the energy used in heating the wire is a loss. In either case the dissipated power is equal to I^2R. In the case of heat losses, R is a real resistance, but in the case of radiation, R is an assumed resistance, which, if present, would dissipate the power actually radiated from the antenna. This assumed resistance is called the *radiation resistance*. The total power loss in the antenna is therefore equal to $I^2 (R_0 + R)$, where R_0 is the radiation resistance and R is the real, or ohmic, resistance.

In the ordinary half wave antenna operated at amateur frequencies, the power lost as heat in the conductor does not exceed a few percent of the total power supplied to the antenna. This is because the RF resistance of copper wire even as small as no. 14 is very low compared with the radiation resistance of an antenna that is reasonably clear of surrounding objects and is not too close to the ground. Therefore it can be assumed that the ohmic loss in a reasonably well-located antenna is negligible, and that all of the resistance shown by the antenna is radiation resistance.

As a radiator of electromagnetic waves, such an antenna is a highly efficient device.

The value of radiation resistance, as measured at the center of a half-wave antenna, depends on a number of factors. One is the location of the antenna with respect to other objects, particularly the earth. Another is the length/diameter ratio of the conductor used. In free space — with the antenna remote from everything else — the radiation resistance of a resonant antenna made of an infinitely thin conductor is approximately 73 ohms. The concept of a free-space antenna forms a convenient basis for calculation because the modifying effect of the ground can be taken into account separately. If the antenna is at least several wavelengths away from ground and other objects, it can be considered to be in free space insofar as its own electrical properties are concerned. This condition can be met easily with antennas in the VHF and UHF range. At these frequencies, antennas are small and a wavelength may be only a few feet (or less) so it is easy to mount the antenna several wavelengths above ground.

As the antenna is made thicker, the radiation resistance decreases. For most wire antennas it is close to 65 ohms. The radiation resistance will usually lie between 55 and 60 ohms for antennas constructed of rod or tubing.

The actual value of the radiation resistance — at least as long as it is 50 ohms or more — has no appreciable effect on the radiation efficiency of the antenna. This is because the ohmic resistance is only on the order of 1 ohm with the conductors used for thick antennas. The ohmic resistance does not become important until the radiation resistance drops to very low values — say less than 10 ohms — as may be the case when several antenna elements are coupled to form an array.

The radiation resistance of a resonant antenna is the "load" for the transmitter or for the RF transmission line connecting the transmitter and antenna. Its value is important, therefore, in determining the way in which the antenna and transmitter or line are coupled. Most modern transmitters require a 50-ohm load. To transfer the maximum amount of power possible, the transmitter output impedance, the transmission-line characteristic impedance and the radiation resistance should all be equal, or matched by means of an appropriate impedance-matching network.

[Now study FCC examination questions with numbers that begin 4AI-4. Review this section as needed.]

ANTENNA EFFICIENCY

The *efficiency of an antenna* is given by:

$$\text{Efficiency} = \frac{(R_R)}{(R_T)} \times 100\% \qquad \qquad \text{(Eq. 9-9)}$$

where R_R is the radiation resistance and R_T is the total resistance. The total resistance includes radiation resistance, resistance in conductors and dielectrics (including the resistance of loading coils, if used), and the resistance of the grounding system, usually referred to as "ground resistance."

It was stated earlier in this chapter that a half-wave antenna operates at very high efficiency because the conductor resistance is negligible compared with the radiation resistance. In the case of the grounded antenna, the ground resistance usually is not negligible, and if the antenna is short (compared with a quarter wavelength) the resistance of the necessary loading coil may become appreciable. To attain an efficiency comparable with that of a half-wave antenna in a grounded one having a height of ¼ wavelength or less, great care must be used to reduce both ground resistance and the resistance of any required loading inductors. Without a fairly elaborate grounding system, the efficiency is not likely to exceed 50 percent and may be much less, particularly at heights below ¼ wavelength. A ¼-wavelength ground-mounted vertical antenna normally includes many radial wires, laid out as spokes on a wheel, with the antenna element in the center.

If a half-wave dipole antenna has a radiation resistance of 70 ohms and a total resistance of 75 ohms, Eq. 9-9 tells us the efficiency of the antenna:

$$\text{Efficiency} = \frac{(70 \text{ ohms})}{(75 \text{ ohms})} \times 100\% = 93\%$$

As another example, let's calculate the efficiency of a ground-mounted vertical antenna that has a radiation resistance of 25 ohms and a total resistance of 70 ohms. Again using Eq. 9-9, we find that the efficiency of this antenna is 36%.

[Before proceeding, study FCC examination questions with numbers that begin 4AI-6. Review this section as needed.]

PARASITIC ELEMENTS

A *parasitic element* receives its excitation from mutual coupling rather than from a transmission line. Parasitic elements are used in directional antennas. At high frequencies, the most common directional antenna is the Yagi beam. It consists of a driven element equal to an electrical half wavelength at the operating frequency. A two-element beam may have a *director*, which is slightly shorter (about 5%) and is placed forward of the driven element. A two-element beam may also have a *reflector* that is slightly longer (about 5%) and is placed behind the driven element. See Fig. 9-7. A three-element beam will have both a director and a reflector. Larger beams have additional directors.

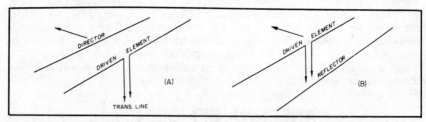

Fig. 9-7 — Antenna systems using a single parasitic element. In A, the parasitic element acts as a director, while in B it serves as a reflector. The arrows show the direction in which maximum radiation takes place.

Reflectors and directors are known as parasitic elements because they are not fed directly from the feed line. Instead, the field generated by the driven element generates currents in them, which in turn generate new fields. Since these fields ideally add together in the forward direction but cancel in the back and side directions, the transmitted energy is concentrated into a beam going out in the forward direction.

[Study FCC examination questions with numbers that begin 4AI-3. Review this section as needed.]

ANTENNA GAIN, BEAMWIDTH AND BANDWIDTH

In a perfect directional antenna, the radio wave would be concentrated in the forward direction only. This is known as the *major lobe of radiation*. (Most beams also have *minor lobes* in the back and side directions.)

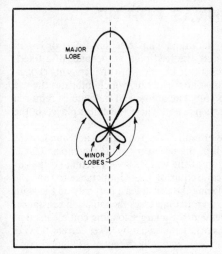

Fig. 9-8 — Radiation pattern for a hypothetical beam antenna, illustrating major and minor radiation lobes.

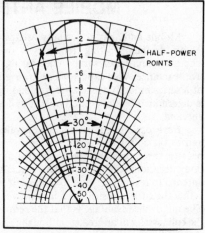

Fig. 9-9 — The width of a beam is the angular distance between the directions at which the received or transmitted power is one half the maximum power (– 3 dB).

Fig. 9-8 is an example of a radiation pattern for a beam antenna, illustrating major and minor pattern lobes. The greater the number of elements and the longer the distance between elements (up to an optimum spacing), the narrower the radiated beam. By reducing radiation in the side and back directions and concentrating it instead into a narrow beam in the forward direction, a beam antenna can have more effective radiated power than a dipole. The ratio (expressed in decibels) between the signal radiated from a beam and the signal radiated from a reference antenna (usually a dipole) at the same transmitting location is called the *gain* of the beam. A typical beam might have 6 dB of gain, which means that it makes your signal sound four times (6 dB) louder than if you were using a dipole with the same transmitter.

The gain of directional antennas is the result of concentrating the radio wave in one direction at the expense of radiation in other directions. Since practical antennas are not perfect, there is always some radiation in undesired directions as well. A plot of relative field strength in all horizontal directions is called the horizontal-radiation pattern. The vertical-radiation pattern is a similar plot of the field strength in the vertical plane.

The *beamwidth* is the angular distance between the points on either side of the main direction, at which the gain is 3 dB below the maximum. See Fig. 9-9. A three-element beam, for example, might be found to have a beamwidth of 74°. This means if you turn your beam plus or minus 37° from the optimum heading, the signal you receive (and the signal received from your transmitter) will drop by 3 dB.

The term *antenna bandwidth* refers generally to the range of frequencies over which an antenna can be used to obtain good performance. The bandwidth is usually referenced to some SWR value, such as, "The 2:1 SWR bandwidth is 3.5 to 3.8 MHz." This means that the SWR between 3.5 and 3.8 MHz will be 2:1 or lower.

[Go to the FCC examination questions with numbers that begin 4AI-1, and study them before proceeding. Review this section as needed.]

MOBILE ANTENNAS FOR HF

Mobile antennas are usually vertically mounted whip antennas 8 feet or less in length. As the operating frequency is lowered, the feed-point impedance of a fixed-length antenna appears to be a decreasing resistance in series with an increasing capacitive reactance. This capacitive reactance must be tuned out, which indicates the use of a series inductive reactance, or loading coil. The amount of inductance required is determined by the desired operating frequency and where the coil is placed in the antenna.

Base loading requires the lowest value of inductance for a given antenna length, and as the coil is moved farther up the whip, the necessary value increases. This is because the capacitance between the portion of the whip above the coil and the car body decreases (higher capacitive reactance), requiring more inductance to tune the antenna to resonance. One advantage of placing the coil at least part way up the whip is that the current distribution is improved, and that increases the radiation resistance. The major disadvantage is that the requirement for a larger loading coil means that the coil losses will be greater, although this is offset somewhat by lower current flowing through the larger coil. *Center loading* has been generally accepted as a good compromise in mobile-antenna design.

Fig. 9-10 shows a typical bumper-mounted, center-loaded whip antenna suitable for operation in the HF range. The antenna could also be mounted on the car body proper (such as a fender). The base spring acts as a shock absorber for the base of the whip, since the continual flexing while the car is in motion would otherwise weaken the antenna. A short, heavy, mast section is mounted between the base spring and loading coil. Some models have a mechanism that allows the antenna to be tipped over for adjustment or for fastening to the roof of the car when not in use.

It is also advisable to extend a couple of guy lines from the base of the loading coil to clips or hooks fastened to the rain trough on the roof of the car. Nylon fishing line of about 40-pound test strength is suitable for this purpose. The guy lines act as safety cords and also reduce the swaying motion of the antenna considerably. The feed line to the transmitter is connected to the bumper and base of the antenna. Good low-resistance connections are important here.

Tune-up of the antenna is usually accomplished by changing the height of the adjustable whip section above the precut loading coil. First, tune the receiver and try to determine where the signals seem to peak up. Once this frequency is found, check the SWR with the transmitter on, seeking the frequency of lowest SWR. Shortening the adjustable section will increase the resonant frequency and making it longer will lower the frequency. It is important that the antenna be 10 feet or more away from surrounding objects such as overhead wires, since considerable detuning can occur. Once you find the setting where the SWR is lowest at the center of the desired operating frequency range, record the length of the adjustable section.

Loading Coils

The difficulty in constructing suitable loading coils increases as the frequency of operation is lowered for typical antenna lengths used in mobile work. Since the required resonating inductance gets larger and the radiation resistance decreases at lower frequencies, most of the power may be dissipated in the coil resistance and in other ohmic losses. This is one reason why it is advisable to buy a commercially made loading coil with the highest power rating possible, even though you may only be considering low-power operation. Percentage-wise, the coil losses in the higher power loading coils are usually less, with subsequent improvement in radiating efficiency, regardless of the power level used. Of course, this same philosophy also applies to homemade loading coils.

The primary goal here is to provide a coil with the highest Q possible. This means

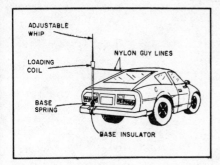

Fig. 9-10 — A typical bumper-mounted HF-mobile antenna. Note the nylon guy lines.

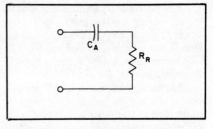

Fig. 9-11 — At frequencies below the resonant frequency, the whip antenna will show capacitive reactance as well as resistance. R_R is the radiation resistance, and C_A represents the capacitive reactance.

the coil should have a high ratio of reactance to resistance, so that heating losses will be minimized. High-Q coils require a large conductor, "air-wound" construction, large spacing between turns, the best insulating material available, a diameter not less than half the length of the coil (this is not always mechanically feasible) and a minimum of metal in the field.

Once the antenna is tuned to resonance, the input impedance at the antenna terminals will look like a pure resistance. Neglecting losses, this value drops from nearly 15 ohms on 15 meters to 0.1 ohm on 160 meters for an 8-foot whip. When coil and other losses are included, the input resistance increases to approximately 20 ohms on 160 meters and 16 ohms on 15 meters. These values are for relatively high-efficiency systems. From this, you can see that the radiating efficiency is much poorer on 160 meters than on 15 meters under typical conditions.

Since most modern gear is designed to operate into a 50-ohm impedance, a matching network may be necessary with some mobile antennas. This can take the form of either a broad-band transformer, a tapped coil or an LC matching network. With homemade or modified designs, the tapped-coil arrangement is perhaps the easiest to build, while the broad-band transformer requires no adjustment. As the losses go up, so does the input resistance, and in less efficient systems the matching network may be eliminated.

The Equivalent Circuit of a Typical Mobile Antenna

Antenna resonance is defined as the frequency at which the input impedance at the antenna terminals is a pure resistance. The shortest length at which this occurs for a vertical antenna over a ground plane is a quarter wavelength at the operating frequency; the impedance value for this length (neglecting losses) is about 36 ohms. The idea of resonance can be extended to antennas shorter (or longer) than a quarter wavelength, and means only that the input impedance is purely resistive. As pointed out previously, when the frequency is lowered, the antenna looks like a series RC circuit, as shown in Fig. 9-11.

The capacitive reactance can be canceled out by connecting an equivalent inductive reactance, L_L, in series as shown in Fig. 9-12. This arrangement tunes the system to resonance at a particular frequency. The price you pay for the shortened antenna is decreased bandwidth.

Mobile Antenna Efficiency

Antenna efficiency was discussed earlier in this chapter. In that section the importance of minimizing ohmic losses was discussed. If you have trouble understanding the material here, review that section.

For lowest loss, an amateur would use a *self-resonant antenna,* such as a dipole or a quarter wavelength vertical. That is not possible, of course, for HF mobile operation, except at the upper end of the range.

Mobile antenna losses at HF, for the most part, are caused by two factors: ground return resistance and loading coil losses. To minimize ground losses, the transmission line should be connected to the metal automobile body through a low resistance and low reactance connection. A good way to do this is with a short length of ground strap or coaxial-cable braid.

Another method that can be used successfully to reduce loading-coil losses is a technique called *top loading.* This method calls for a "capacitive hat" to be added above the loading coil, either just above the coil or near the top of the antenna whip.

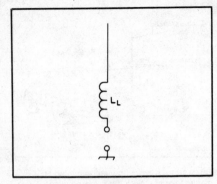

Fig. 9-12 — At frequencies lower than the resonant frequency, the capacitive reactance of a whip antenna can be canceled by adding an equivalent inductive reactance, in the form of a loading coil, in series with the antenna.

The added capacitance at the top of the whip allows a smaller value of load inductance. This reduces the amount of loss in the system, and improves the antenna radiation efficiency.

[At this point, you should go to Chapter 10, and study those FCC examination questions with numbers that begin 4AI-11 and 4A1-12. Review this section as needed.]

Element 4A Question Pool

Don't Start Here!

B efore you read the questions and answers printed here, be sure to read the appropriate text in the previous chapters. Use these questions as review exercises, when suggested in the text. You should not attempt to memorize all 507 questions and answers. The material presented in this book has been carefully written and scientfically prepared to guide you step by step though the learning process. By understanding the electronics principles and Amateur Radio concepts as they are presented, your insight into our hobby and your appreciation for the privileges granted by each license class will be greatly enhanced.

This chapter contains the complete FCC Question Pool for the Advanced class exam, Element 4A. It includes questions that have been written by amateurs and submitted to the FCC. The FCC has edited the questions, and released this set of 507 in January 1986. The question numbers are keyed to specific items on the Element 4A Syllabus, printed at the end of Chapter 1 in this book.

Your Element 4A exam will consist of 50 of these questions, to be selected by the Volunteer Exam Coordinator that is overseeing the test session. The multiple-choice answers and distractors presented here, along with an answer key, have been carefully prepared and evaluated by the ARRL Staff and volunteers in the Field Organization of the League. The FCC specifies how many questions from each section must be on your test. For example, there must be six questions from the Rules and Regulations section, subelement 4AA; one question from the Operating Procedures section, subelement 4AB; and so on. Table 10-1 shows the number of questions from each subelement that will be on your exam. The number of questions to be selected from each section is printed at the beginning of each subelement.

The FCC requires that the questions used on the exam be used word-for-word, exactly as released by them and printed in PR Bulletin 1035C. When a Volunteer Examiner Coordinator selects a question from the bulletin for use on an exam, it must be used exactly as printed. VECs are not at liberty to correct spelling mistakes or other perceived typographical errors. They may not delete superfluous words or insert obviously missing text. To make your preparation for an examination easier and less confusing, we have marked with an asterisk those questions about which our editors had any doubt. We have also tried to word the answers to these questions in such a way that the answers help complete the questions.

To repeat, then, the actual questions printed here are exactly as released by the FCC, and must be used without change on your exam. The multiple-choice answers included with the questions have been released by the ARRL/VEC Office to all VECs and to all publishers of Amateur Radio study materials. Many VECs, including the ARRL/VEC, will be using these answers in conjunction with the question on your exam.

Table 10-1
Advanced Class Exam Content

Subelement	4AA	4AB	4AC	4AD	4AE	4AF	4AG	4AH	4AI
Number of questions	6	1	2	4	10	6	10	6	5

SUBELEMENT 4A—Rules and Regulations (6 questions)

4AA-1.1 What are the frequency privileges authorized to the Advanced operator in the 75 meter band?
A. 3525 kHz to 3750 kHz and 3775 kHz to 4000 kHz
B. 3500 kHz to 3525 kHz and 3800 kHz to 4000 kHz
C. 3500 kHz to 3525 kHz and 3800 kHz to 3890 kHz
D. 3525 kHz to 3775 kHz and 3800 kHz to 4000 kHz

4AA-1.2 What are the frequency privileges authorized to the Advanced operator in the 40 meter band?
A. 7000 kHz to 7300 kHz
B. 7025 kHz to 7300 kHz
C. 7025 kHz to 7350 kHz
D. 7000 kHz to 7025 kHz

4AA-1.3 What are the frequency privileges authorized to the Advanced operator in the 20 meter band?
A. 14000 kHz to 14150 kHz and 14175 kHz to 14350 kHz
B. 14025 kHz to 14175 kHz and 14200 kHz to 14350 kHz
C. 14000 kHz to 14025 kHz and 14200 kHz to 14350 kHz
D. 14025 kHz to 14150 kHz and 14175 kHz to 14350 kHz

4AA-1.4 What are the frequency privileges authorized to the Advanced operator in the 15 meter band?
A. 21000 kHz to 21200 kHz and 21250 kHz to 21450 kHz
B. 21000 kHz to 21200 kHz and 21300 kHz to 21450 kHz
C. 21025 kHz to 21200 kHz and 21225 kHz to 21450 kHz
D. 21025 kHz to 21250 kHz and 21270 kHz to 21450 kHz

4AA-2.1 What is meant by automatic retransmission?
A. The retransmitting station is actuated by a received electrical signal
B. The retransmitting station is actuated by a telephone control link
C. The retransmitting station is actuated by a control operator
D. The retransmitting station is actuated by a call sign sent in Morse code

4AA-2.2 What is the term for the retransmission of signals by an amateur radio station whereby the retransmitting station is actuated solely by the presence of a received signal through electrical or electromechanical means, i.e., without any direct, positive action by the control operator?
A. Simplex retransmission
B. Manual retransmission
C. Linear retransmission
D. Automatic retransmission

4AA-2.3 Under what circumstances, if any, may an amateur station automatically retransmit programs or the radio signals of other amateur stations?
A. Only when the station licensee is present
B. Only when in repeater operation
C. Only when the control operator is present
D. Only during portable operation

4AA-2.4 What is meant by manual retransmission?
A. A retransmitted signal that is not automatically controlled
B. A retransmit signal that is automatically controlled
C. An OSCAR satellite transponder
D. The theory behind operational repeaters

4AA-3.1   What is meant by repeater operation?
    A. An amateur radio station employing a phone patch to pass third party traffic
    B. An apparatus for effecting remote control between a control point and a remotely controlled station
    C. Manual or simplex operation
    D. Radio communications in which amateur radio station signals are automatically retransmitted

4AA-3.2   What is a closed repeater?
    A. A repeater containing control circuitry that limits access to the repeater to members of a certain group
    B. A repeater containing no special control circuitry to limit access to any licensed amateur
    C. A repeater containing a transmitter and receiver on the same frequency, a closed pair
    D. A repeater shut down by order of an FCC District Engineer-in-Charge

4AA-3.3   What frequencies in the 10 meter band are available for repeater operation?
    A. 28.0-28.7 MHz
    B. 29.0-29.7 MHz
    C. 29.5-29.7 MHz
    D. 28.5-29.7 MHz

4AA-3.4   What determines the maximum effective radiated power a station in repeater operation may use?
    A. Repeaters are authorized 1500 watts power output at all times
    B. The percent modulation and emission type used
    C. Polarization and direction of major lobes
    D. Frequency and antenna height above average terrain

4AA-3.5   How is effective radiated power determined?
    A. By measuring the output power of the final amplifier
    B. By dividing the final amplifier power by the feed-line losses
    C. By calculating the product of the transmitter power to the antenna and the antenna gain
    D. By measuring the power delivered to the antenna

4AA-3.6   What is an open repeater?
    A. A repeater that contains no special control circuitry to limit access to any licensed amateur
    B. A repeater available for use only by members of a club or repeater group
    C. A repeater that continuously transmits a signal to indicate that it is available for use
    D. A repeater whose frequency pair has been properly coordinated

4AA-3.7   What frequencies in the 6 meter band are available for repeater operation?
    A. 51.00-52.00 MHz
    B. 50.25-52.00 MHz
    C. 52.00-53.00 MHz
    D. 52.00-54.00 MHz

4AA-3.8   What frequencies in the 2 meter band are available for repeater operation?
   A. 144.50-145.50 and 146-148.00 MHz
   B. 144.50-148.00 MHz
   C. 144.75-146.00 and 146-148.00 MHz
   D. 146.00-148.00 MHz

4AA-3.9   What frequencies in the 1.25 meter band are available for repeater operation?
   A. 220.25-225.00 MHz
   B. 220.50-225.00 MHz
   C. 221.00-225.00 MHz
   D. 223.00-225.00 MHz

4AA-3.10   What frequencies in the 0.70 meter band are available for repeater operation?
   A. 420.0-431, 433-435 and 438-450 MHz
   B. 420.5-440 and 445-450 MHz
   C. 420.5-435 and 438-450 MHz
   D. 420.5-433, 435-438 and 439-450 MHz

4AA-4.1   What is meant by auxiliary operation?
   A. Radio communication from a location more than 50 miles from that indicated on the station license for a period of more than three months
   B. Remote control of model airplanes or boats using frequencies above 50.1 MHz
   C. Remote control of model airplanes or boats using frequencies above 29.5 MHz
   D. Radio communications for remotely controlling other amateur radio stations, for automatically relaying the signals of other amateur stations in a system of stations or for intercommunicating with other amateur stations in a system of stations

4AA-4.2   What are three uses for stations in auxiliary operation?
   A. Remote control of other amateur stations, automatically relaying signals of other amateur stations in a system of stations and intercommunicating with other amateur stations in a system of amateur radio stations
   B. Remote control of model craft and vehicles, automatically relaying signals of other amateur stations in a system of stations and intercommunicating with other amateur stations in a system of stations
   C. Remote control of other amateur stations and of model craft and vehicles, manually relaying signals of other amateur stations in a system of stations and intercommunicating with other amateur stations in a system of amateur radio stations
   D. Operation for more than three months at a location more than 50 miles from the location listed on the station license, automatically relaying signals from other amateur stations in a system of stations and intercommunicating with other amateur stations in a system of amateur radio stations

4AA-4.3   A station in auxiliary operation may only communicate with which stations?
   A. Stations in the public safety service
   B. Other amateur stations in the system of amateur stations shown on the system network diagram
   C. Amateur radio stations in space satellite operation
   D. Amateur radio stations other than those under manual control

4AA-4.4 What frequencies are authorized for stations in auxiliary operation?
A. All amateur frequency bands above 220.5 MHz, except 432-433 MHz and 436-438 MHz
B. All amateur frequency bands above 220.5 MHz, except 431-432 MHz and 435-437 MHz
C. All amateur frequency bands above 220.5 MHz, except 431-433 MHz and 435-438 MHz
D. All amateur frequency bands above 220.5 MHz, except 430-432 MHz and 434-437 MHz

4AA-5.1 What is meant by remote control of an amateur radio station?
A. Amateur communications conducted from a specific geographical location other than that shown on the station license
B. Automatic operation of a station from a control point located elsewhere than at the station transmitter
C. An amateur radio station operating under automatic control
D. Manual operation of a station from a control point located elsewhere than at the station transmitter

4AA-5.2 How do the responsibilities of the control operator of a station under remote control differ from one under local control?
A. Provisions must be made to limit transmissions to no more than 3 minutes if the control link malfunctions
B. Provisions must be made to limit transmissions to no more than 4 minutes if the control link malfunctions
C. Provisions must be made to limit transmissions to no more than 5 minutes if the control link malfunctions
D. Provisions must be made to limit transmissions to no more than 10 minutes if the control link malfunctions

4AA-5.3 If the control link for a station under remote control malfunctions, how long may the station continue to transmit?
A. 5 seconds
B. 10 minutes
C. 3 minutes
D. 5 minutes

4AA-5.4 What frequencies are authorized for radio remote control of an amateur radio station?
A. All amateur frequency bands above 220.5 MHz, except 432-433 MHz and 436-438 MHz
B. All amateur frequency bands above 220.5 MHz, except 431-432 MHz and 435-437 MHz
C. All amateur frequency bands above 220.5 MHz, except 431-433 MHz and 435-438 MHz
D. All amateur frequency bands above 220.5 MHz, except 430-432 MHz and 434-437 MHz

4AA-5.5 What frequencies are authorized for radio remote control of a station in repeater operation?
A. All amateur frequency bands above 220.5 MHz, except 432-433 MHz and 436-438 MHz
B. All amateur frequency bands above 220.5 MHz, except 431-432 MHz and 435-437 MHz
C. All amateur frequency bands above 220.5 MHz, except 430-432 MHz and 434-437 MHz
D. All amateur frequency bands above 220.5 MHz, except 431-433 MHz and 435-438 MHz

*4AA-6.1   What is meant by <u>automatic control</u> of an amateur
           A. Automatic control of an Amateur Radio station is the use of
              devices and procedures for control so that a control operator
              does not have to be present at the control point at all times
           B. Automatic control of an Amateur Radio station is radio communi-
              cation for remotely controlling another amateur radio station
           C. Automatic control of an Amateur Radio station is remotely con-
              trolling a station such that a control operator does not have to
              be present at the control point at all times
           D. Automatic control of an Amateur Radio station is the use of a
              control link between a control point and a remotely controlled
              station

4AA-6.2   How do the responsibilities of the control operator of a station under
           automatic control differ from one under local control?
           A. Under local control, there is no control operator
           B. Under automatic control, a control operator is not required to be
              present at the control point at all times
           C. Under automatic control, there is no control operator
           D. Under local control, a control operator is not required to be
              present at the control point at all times

4AA-6.3   Which amateur stations may be operated by automatic control?
           A. Stations without a control operator
           B. Stations in repeater operation
           C. Stations that do not have transmission-limiting timing devices
           D. Stations that transmit codes and cipher groups, as defined in
              FCC Section 97.117

4AA-7.1   What is a <u>control link</u>?
           A. The automatic control devices of an unattended station
           B. An automatically operated link
           C. The remote control apparatus between a control point and a
              remotely controlled station
           D. A transmission-limiting timing device

4AA-7.2   What is the term for apparatus to effect remote control between the
           control point and a remotely controlled station?
           A. Tone link
           B. Wire control
           C. Remote control
           D. Control link

4AA-8.1   What is a <u>system network diagram</u>?
           A. As defined in Section 97.3, a diagram showing each station in a
              system of stations, and its relationship to other stations and to
              the control point
           B. As defined in Section 97.3, a diagram describing a computer
              interface to an amateur radio station
           C. As defined in Section 97.3, a diagram demonstrating how
              mobile amateur radio station used on board a ship or aircraft is
              electrically separate from and independent of all other radio
              equipment on board
           D. As defined in Section 97.3, a diagram showing the stages of an
              amateur transmitter or external radio frequency power amplifier

4AA-8.2 What type of diagram shows each station and its relationship to other stations in a network of amateur stations, and to the control point(s)?
- A. A control link diagram
- B. A system network diagram
- C. A radio network diagram
- D. A control point diagram

4AA-9.1 At what level of modulation must an amateur station in repeater operation transmit its identification?
- A. At a level sufficient to completely block the repeated transmission
- B. At a level low enough to cause no interference to users of the repeater
- C. At a level sufficient to be intelligible through the repeated transmission
- D. At a 150% modulation level, as required by Section 97.84

4AA-9.2 At what level of modulation must an amateur station in auxiliary operation transmit its identification?
- A. At a level sufficient to completely block the repeated transmission
- B. At a level low enough to cause no interference to users of the repeater
- C. At a level sufficient to be intelligible through the repeated transmission
- D. At a 150% modulation level, as required by Section 97.84

4AA-9.3 What additional station identification requirements apply to amateur stations in repeater operation?
- A. The letters "AUX" must follow the station call sign when identifying by radiotelegraphy
- B. The letters "RPTR" must follow the station call sign when identifying by radiotelegraphy
- C. The word "auxiliary" must be added after the call sign when identifying by radiotelephony
- D. The word "repeater" must be added after the call sign when identifying by radiotelephony

4AA-9.4 What additional station identification requirements apply to amateur stations in auxiliary operation?
- A. The word "auxiliary" must be transmitted at the end of the call sign when identifying by radiotelephony
- B. The letters "RPTR" must precede the station call sign when identifying by radiotelegraphy
- C. The letters "AUX" must precede the station call sign when identifying by radiotelegraphy
- D. The words "remote control" must be added after the call sign when identifying by radiotelephony

4AA-10.1 When is prior FCC approval required before constructing or altering an amateur station antenna structure?
- A. When the antenna structure violates local building codes
- B. When the height above ground will exceed 200 feet
- C. When an antenna located 23000 feet from an airport runway will be 150 feet high
- D. When an antenna located 23000 feet from an airport runway will be 100 feet high

4AA-10.2 What must an amateur radio operator obtain from the FCC before constructing or altering an antenna structure more than 200 feet high?
   A. An Environmental Impact Statement
   B. A Special Temporary Authorization
   C. Prior approval
   D. An effective radiated power statement

4AA-11.1 How is antenna height above average terrain determined?
   A. By an aerial survey
   B. The height of the center of radiation of the antenna above an averaged value of the elevation above sea level for surrounding terrain
   C. The height of the antenna above the highest value of the elevation above sea level for surrounding terrain
   D. By measuring the highest point of the antenna above the lowest value of surrounding terrain

4AA-11.2 For a station in repeater operation transmitting on 146.94 MHz, what is the maximum ERP permitted for an antenna height above average terrain of more than 1050 feet?
   A. 100 watts
   B. 200 watts
   C. 400 watts
   D. 800 watts

4AA-12.1 What are business communications?
   A. Third party traffic that involves material compensation
   B. Any transmission that facilitates the regular business or commercial affairs of any party
   C. Transmissions ensuring safety on a highway, such as calling a commercial tow truck service
   D. An autopatch using a commercial telephone system

4AA-12.2 What is the term for a transmission or communication the purpose of which is to facilitate the regular business or commercial affairs of any party?
   A. Duplex autopatch
   B. Third party traffic that involves compensation
   C. Business communications
   D. Simplex autopatch

4AA-12.3 Under what conditions, if any, may business communications be transmitted by an amateur station?
   A. When the total remuneration does not exceed $25
   B. When the control operator is employed by the FCC
   C. When transmitting international third party traffic
   D. During an emergency

4AA-13.1 What are the only types of messages that may be transmitted to an amateur station in a foreign country?
   A. Call sign and signal reports
   B. Emergency messages
   C. Business messages
   D. Personal remarks

4AA-13.2 What are the limitations on international amateur radiocommunications regarding the types of messages transmitted?
   A. Emergency communications only
   B. Technical or personal messages only
   C. Business communications only
   D. Call sign and signal reports only

4AA-14.1   Under what circumstances, if any, may amateur operators accept payment for using their stations to send messages?
  A. When employed by the FCC
  B. When passing emergency traffic
  C. Under no circumstances
  D. When passing international third party traffic

4AA-14.2   Under what circumstances, if any, may the licensee of an amateur station in repeater operation accept remuneration for providing communication services to another party?
  A. When the repeater is operating under portable power
  B. When the repeater is under local control
  C. During Red Cross or other emergency service drills
  D. Under no circumstances

4AA-15.1   Who is responsible for preparing an Element 1(A) telegraphy examination?
  A. The examiner
  B. The FCC
  C. The VEC
  D. Any Novice licensee

4AA-15.2   What must the Element 1(A) telegraphy examination prove?
  A. The applicant's ability to send and receive text in international Morse code at a rate of not less than 13 words per minute
  B. The applicant's ability to send and receive text in international Morse code at a rate of not less than 5 words per minute
  C. The applicant's ability to send and receive text in international Morse code at a rate of not less than 20 words per minute
  D. The applicant's ability to send text in international Morse code at a rate of not less than 13 words per minute

4AA-15.3   Which telegraphy characters are used in an Element 1(A) telegraphy examination?
  A. The letters A through Z, Ø through 9, the period, the comma, the question mark, AR, SK, BT and DN
  B. The letters A through Z, Ø through 9, the period, the comma, the open and closed parenthesis, the question mark, AR, SK, BT and DN
  C. The letters A through Z, Ø through 9, the period, the comma, the dollar sign, the question mark, AR, SK, BT and DN
  D. A through Z, Ø through 9, the period, the comma, and the question mark

4AA-16.1   Who is responsible for preparing an Element 2 written examination?
  A. The FCC
  B. Any Novice licensee
  C. The test examiner
  D. The VEC

4AA-16.2   Where do volunteer examiners obtain the questions for preparing an Element 2 written examination?
  A. From FCC PR Bulletin 1035C
  B. From FCC PR Bulletin 1035B
  C. From FCC PR Bulletin 1035D
  D. From FCC PR Bulletin 1035A

4AA-17.1 Who is eligible for administering an examination for the Novice operator license?
  A. An amateur radio operator holding a General, Advanced or Extra class license and at least 18 years old
  B. An amateur radio operator holding a Technician, General, Advanced or Extra class license and at least 18 years old
  C. An amateur radio operator holding a General, Advanced or Extra class license and at least 16 years old
  D. An amateur radio operator holding a Technician, General, Advanced or Extra class license and at least 16 years old

4AA-17.2 For how long must the volunteer examiner for a Novice operator examination retain the test papers?
  A. Ten years from the date of the examination
  B. One year from the date of the examination
  C. Twelve years from the date of the examination
  D. Until the license is issued

4AA-17.3 Where must the volunteer examiner for a Novice operator examination retain the test papers?
  A. With the examinee's station records
  B. With the VEC that issued the papers
  C. With the volunteer examiner's station records
  D. With the Volunteer Examiner Team Chief's station records

4AA-18.1 What is the minimum passing score on a written examination element for the Novice operator license?
  A. 84 percent, minimum
  B. 74 percent, minimum
  C. 70 percent, minimum
  D. 80 percent, minimum

4AA-18.2 For a 20 question Element 2 written examination, how many correct answers constitute a passing score?
  A. 10 or more
  B. 12 or more
  C. 14 or more
  D. 15 or more

4AA-18.3 In a telegraphy examination, how many characters are counted as one word?
  A. 2
  B. 5
  C. 8
  D. 10

4AA-19.1 What is the minimum age to be a volunteer examiner?
  A. 16 years old
  B. 21 years old
  C. 18 years old
  D. 13 years old

4AA-19.2 Under what circumstances, if any, may volunteer examiners be compensated for their services?
  A. Under no circumstances
  B. When out-of-pocket expenses exceed $25
  C. The volunteer examiner may be compensated when traveling over 25 miles to the test site
  D. Only when there are more than 20 applicants attending the examination session

4AA-19.3   Under what circumstances, if any, may a person whose amateur station license or amateur operator license has ever been revoked or suspended be a volunteer examiner?
A. Under no circumstances
B. Only if five or more years have elapsed since the revocation or suspension
C. Only if 3 or more years have elapsed since the revocation of suspension
D. Only after review and subsequent approval by the VEC

4AA-19.4   Under what circumstances, if any, may an employee of a company which is engaged in the distribution of equipment used in connection with amateur radio transmissions be a volunteer examiner?
A. If the employee is employed in the amateur radio sales part of the company
B. If the employee does not normally communicate with the manufacturing or distribution part of the company
C. If the employee serves as a volunteer examiner for his/her customers
D. If the employee does not normally communicate with the benefits and policies part of the company

4AA-20.1   What are the penalties for fraudulently administering examinations?
A. The examiner's station license may be suspended for a period not to exceed 3 months
B. A monetary fine not to exceed $500 for each day the offense was committed
C. Possible revocation of his/her amateur radio station license
D. The examiner may be restricted to giving only Novice class exams

4AA-20.2   What are the penalties for administering examinations for money or other considerations?
A. The examiner's station license may be suspended for a period not to exceed 3 months
B. A monetary fine not to exceed $500 for each day the offense was committed
C. The examiner may be restricted to administering only Novice class license exams
D. Possible revocation of his/her amateur radio station license

SUBELEMENT 4AB—Operating Procedures (1 question)

4AB-1.1   What is <u>facsimile</u>?
           A. The transmission of characters by radioteletype that form a
              picture when printed
           B. The transmission of still pictures by slow-scan television
           C. The transmission of video by amateur television
           D. The transmission of printed pictures for permanent display
              on paper

4AB-1.2   What is the modern standard scan rate for a facsimile picture
           transmitted by an amateur station?
           A. The modern standard is 240 lines per minute
           B. The modern standard is 50 lines per minute
           C. The modern standard is 150 lines per second
           D. The modern standard is 60 lines per second

4AB-1.3   What is the approximate transmission time for a facsimile picture
           transmitted by an amateur station?
           A. Approximately 6 minutes per frame at 240 lpm
           B. Approximately 3.3 minutes per frame at 240 lpm
           C. Approximately 6 seconds per frame at 240 lpm
           D. 1/60 second per frame at 240 lpm

4AB-1.4   What is the term for the transmission of printed pictures by radio?
           A. Television
           B. Facsimile
           C. Xerography
           D. ACSSB

4AB-1.5   In facsimile, how are variations in picture brightness and darkness
           converted into voltage variations?
           A. With an LED
           B. With a Hall-effect transistor
           C. With a photodetector
           D. With an optoisolator

4AB-2.1   What is <u>slow-scan</u> television?
           A. The transmission of Baudot or ASCII signals by radio
           B. The transmission of pictures for permanent display on paper
           C. The transmission of moving pictures by radio
           D. The transmission of still pictures by radio

4AB-2.2   What is the scan rate commonly used for amateur slow-scan
           television?
           A. 20 lines per minute
           B. 15 lines per second
           C. 4 lines per minute
           D. 240 lines per minute

4AB-2.3   How many lines are there in each frame of an amateur slow-scan
           television picture?
           A. 30
           B. 60
           C. 120
           D. 180

4AB-2.4 What is the audio frequency for black in an amateur slow-scan television picture?
- A. 2300 Hz
- B. 2000 Hz
- C. 1500 Hz
- D. 120 Hz

4AB-2.5 What is the audio frequency for white in an amateur slow-scan television picture?
- A. 120 Hz
- B. 1500 Hz
- C. 2000 Hz
- D. 2300 Hz

SUBELEMENT 4AC — Radio Wave Propagation (2 questions)

4AC-1.1   What is a sporadic-E condition?
      A. Variations in E-layer height caused by sunspot variations
      B. A brief increase in VHF signal levels from meteor trails at E-layer height
      C. Patches of dense ionization at E-layer height
      D. Partial tropospheric ducting at E-layer height

*4AC-1.2   What is the propagation condition called where scattered patches of relatively dense ionization develops seasonally at E layer heights?
      A. Auroral propagation
      B. Ducting
      C. Scatter
      D. Sporadic-E

4AC-1.3   In what region of the world is sporadic-E most prevalent?
      A. The equatorial regions
      B. The arctic regions
      C. The northern hemisphere
      D. The polar regions

*4AC-1.4   On which amateur frequency band is extended distant propagation effect of sporadic-E most often observed?
      A. 2 meters
      B. 6 meters
      C. 20 meters
      D. 160 meters

4AC-1.5   What appears to be the major cause of the sporadic-E condition?
      A. Wind shear
      B. Sunspots
      C. Temperature inversions
      D. Meteors

4AC-2.1   What is a selective fading effect?
      A. A fading effect caused by small changes in beam heading at the receiving station
      B. A fading effect caused by phase differences between radio wave components of the same transmission, as experienced atthe receiving station
      C. A fading effect caused by large changes in the height of the ionosphere, as experienced at the receiving station
      D. A fading effect caused by time differences between the receiving and transmitting stations

4AC-2.2   What is the propagation effect called when phase differences between radio wave components of the same transmission are experienced at the recovery station?
      A. Faraday rotation
      B. Diversity reception
      C. Selective fading
      D. Phase shift

4AC-2.3   What is the major cause of selective fading?
           A. Small changes in beam heading at the receiving station
           B. Large changes in the height of the ionosphere, as experi-
              enced at the receiving station
           C. Time differences between the receiving and transmitting
              stations
           D. Phase differences between radio wave components of the
              same transmission, as experienced at the receiving station

4AC-2.4   Which emission modes suffer the most from selective fading?
           A. CW and SSB
           B. FM and double sideband AM
           C. SSB and AMTOR
           D. SSTV and CW

4AC-2.5   How does the bandwidth of the transmitted signal affect selective
           fading?
           A. It is more pronounced at wide bandwidths
           B. It is more pronounced at narrow bandwidths
           C. It is equally pronounced at both narrow and wide bandwidths
           D. The receiver bandwidth determines the selective fading
              effect

4AC-3.1   What effect does auroral activity have upon radio communica-
           tions?
           A. The readability of SSB signals increases
           B. FM communications are clearer
           C. CW signals have a clearer tone
           D. CW signals have a fluttery tone

4AC-3.2   What is the cause of auroral activity?
           A. A high sunspot level
           B. A low sunspot level
           C. The emission of charged particles from the sun
           D. Meteor showers concentrated in the northern latitudes

4AC-3.3   In the northern hemisphere, in which direction should a direc-
           tional antenna be pointed to take maximum advantage of auroral
           propagation?
           A. South
           B. North
           C. East
           D. West

4AC-3.4   Where in the ionosphere does auroral activity occur?
           A. At F-layer height
           B. In the equatorial band
           C. At D-layer height
           D. At E-layer height

4AC-3.5   Which emission modes are best for auroral propagation?
           A. CW and SSB
           B. SSB and FM
           C. FM and CW
           D. RTTY and AM

4AC-4.1   Why does the radio-path horizon distance exceed the geometric horizon?
     A. E-layer skip
     B. D-layer skip
     C. Auroral skip
     D. Radio waves may be bent

4AC-4.2   How much farther does the radio-path horizon distance exceed the geometric horizon?
     A. By approximately 1/3 the distance
     B. By approximately twice the distance
     C. By approximately one-half the distance
     D. By approximately four times the distance

4AC-4.3   To what distance is VHF propagation ordinarily limited?
     A. Approximately 1000 miles
     B. Approximately 500 miles
     C. Approximately 1500 miles
     D. Approximately 2000 miles

4AC-4.4   What propagation condition is usually indicated when a VHF signal is received from a station over 500 miles away?
     A. D-layer absorption
     B. Faraday rotation
     C. Tropospheric ducting
     D. Moonbounce

4AC-4.5   What happens to a radio wave as it travels in space and collides with other particles?
     A. Kinetic energy is given up by the radio wave
     B. Kinetic energy is gained by the radio wave
     C. Aurora is created
     D. Nothing happens since radio waves have no physical substance

SUBELEMENT 4AD — Amateur Radio Practice (4 questions)

4AD-1.1   What is a frequency standard?
　　　　　A. A net frequency
　　　　　B. A device used to produce a highly accurate reference frequency
　　　　　C. A device for accurately measuring frequency to within 1 Hz
　　　　　D. A device used to generate wideband random frequencies

4AD-1.2   What is a frequency-marker generator?
　　　　　A. A device used to produce a highly accurate reference frequency
　　　　　B. A sweep generator
　　　　　C. A broadband white noise generator
　　　　　D. A device used to generate wideband random frequencies

4AD-1.3   How is a frequency-marker generator used?
　　　　　A. In conjunction with a grid-dip meter
　　　　　B. To provide reference points on a receiver dial
　　　　　C. As the basic frequency element of a transmitter
　　　　　D. To directly measure wavelength

4AD-1.4   What is a frequency counter?
　　　　　A. A frequency measuring device
　　　　　B. A frequency marker generator
　　　　　C. A device that determines whether or not a given frequency is in use before automatic transmissions are made
　　　　　D. A broadband white noise generator

4AD-1.5   How is a frequency counter used?
　　　　　A. To provide reference points on an analog receiver dial
　　　　　B. To generate a frequency standard
　　　　　C. To measure the deviation in an FM transmitter
　　　　　D. To measure frequency

4AD-1.6   What is the most the actual transmitter frequency could differ from a reading of 146,520,000-Hertz on a frequency counter with a time base accuracy of $+/-1.0$ ppm?
　　　　　A. 165.2 Hz
　　　　　B. 14.652 kHz
　　　　　C. 146.52 Hz
　　　　　D. 1.4652 MHz

4AD-1.7   What is the most the actual transmitter frequency could differ from a reading of 146,520,000-Hertz on a frequency counter with a time base accuracy of $+/-0.1$ ppm?
　　　　　A. 14.652 Hz
　　　　　B. 0.1 MHz
　　　　　C. 1.4652 Hz
　　　　　D. 1.4652 kHz

4AD-1.8   What is the most the actual transmitter frequency could differ from a reading of 146,520,000-Hertz on a frequency counter with a time base accuracy of $+/-10$ ppm?
　　　　　A. 146.52 Hz
　　　　　B. 10 Hz
　　　　　C. 146.52 kHz
　　　　　D. 1465.20 Hz

4AD-1.9   What is the most the actual transmitter frequency could differ from a
           reading of 432,100,000-Hertz on a frequency counter with a time base
           accuracy of +/- 1.0 ppm?
           A. 43.21 MHz
           B. 10 Hz
           C. 1.0 MHz
           D. 432.1 Hz

4AD-1.10   What is the most the actual transmit frequency could differ from a
           reading of 432,100,000-Hertz on a frequency counter with a time base
           accuracy of +/- 0.1 ppm?
           A. 43.21 Hz
           B. 0.1 MHz
           C. 432.1 Hz
           D. 0.2 MHz

4AD-1.11   What is the most the actual transmit frequency could differ from a
           reading of 432,100,000-Hertz on a frequency counter with a time base
           accuracy of +/- 10 ppm?
           A. 10 MHz
           B. 10 Hz
           C. 4321 Hz
           D. 432.1 Hz

4AD-2.1   What is a dip-meter?
           A. A field strength meter
           B. An SWR meter
           C. A variable LC oscillator with metered feedback current
           D. A marker generator

4AD-2.2   Why is a dip-meter used by many amateur operators?
           A. It can measure signal strength accurately
           B. It can measure frequency accurately
           C. It can measure transmitter output power accurately
           D. It can give an indication of the resonant frequency of a circuit

4AD-2.3   How does a dip-meter function?
           A. Reflected waves at a specific frequency desensitize the detector
              coil
           B. Power coupled from an oscillator causes a decrease in metered
              current
           C. Power from a transmitter cancels feedback current
           D. Harmonics of the oscillator cause an increase in resonant circuit
              Q

4AD-2.4   What two ways could a dip-meter be used in an amateur station?
           A. To measure resonant frequency of antenna traps and to
              measure percentage of modulation
           B. To measure antenna resonance and to measure percentage of
              modulation
           C. To measure antenna resonance and to measure antenna
              impedance
           D. To measure resonant frequency of antenna traps and to
              measure a tuned circuit resonant frequency

4AD-2.5   What types of coupling occur between a dip-meter and a tuned circuit
           being checked?
           A. Resistive and inductive
           B. Inductive and capacitive
           C. Resistive and capacitive
           D. Strong field

4AD-2.6   How tight should the dip-meter be coupled with the tuned circuit being checked?
A. As loosely as possible, for best accuracy
B. As tightly as possible, for best accuracy
C. First loose, then tight, for best accuracy
D. With a soldered jumper wire between the meter and the circuit
D. to be checked, for best accuracy

4AD-2.7   What happens in a dip-meter when it is too tightly coupled with the tuned circuit being checked?
A. Harmonics are generated
B. A less accurate reading results
C. Cross modulation occurs
D. Intermodulation distortion occurs

4AD-3.1   What factors limit the accuracy, frequency response, and stability of an oscilloscope?
A. Sweep oscillator quality and deflection amplifier bandwidth
B. Tube face voltage increments and deflection amplifier voltage
C. Sweep oscillator quality and tube face voltage increments
D. Deflection amplifier output impedance and tube face frequency increments

4AD-3.2   What factors limit the accuracy, frequency response, and stability of a D'Arsonval movement type meter?
A. Calibration, coil impedance and meter size
B. Calibration, series resistance and electromagnet current
C. Coil impedance, electromagnet voltage and movement mass
D. Calibration, mechanical tolerance and coil impedance

4AD-3.3   What factors limit the accuracy, frequency response, and stability of a frequency counter?
A. Number of digits in the readout, speed of the logic and time base stability
B. Time base accuracy, speed of the logic and time base stability
C. Time base accuracy, temperature coefficient of the logic and time base stability
D. Number of digits in the readout, external frequency reference and temperature coefficient of the logic

4AD-3.4   How can the frequency response of an oscilloscope be improved?
A. By using a triggered sweep and a crystal oscillator as the time base
B. By using a crystal oscillator as the time base and increasing the vertical sweep rate
C. By increasing the vertical sweep rate and the horizontal amplifier frequency response
D. By increasing the horizontal sweep rate and the vertical amplifier frequency response

4AD-3.5   How can the accuracy of a frequency counter be improved?
A. By using slower digital logic
B. By improving the accuracy of the frequency response
C. By increasing the accuracy of the time base
D. By using faster digital logic

4AD-4.1 What is the condition called which occurs when the signals of two transmitters in close proximity mix together in one or both of their final amplifiers, and unwanted signals at the sum and difference frequencies of the original transmissions are generated?
A. Amplifier desensitization
B. Neutralization
C. Adjacent channel interference
D. Intermodulation interference

4AD-4.2 How does <u>intermodulation</u> <u>interference</u> between two transmitters usually occur?
A. When the signals from the transmitters are reflected out of phase from airplanes passing overhead
B. When they are in close proximity and the signals mix in one or both of their final amplifiers
C. When they are in close proximity and the signals cause feed back in one or both of their final amplifiers
D. When the signals from the transmitters are reflected in phase from airplanes passing overhead

4AD-4.3 How can intermodulation interference between two transmitters in close proximity often be reduced or eliminated?
A. By using a Class C final amplifier with high driving power
B. By installing a terminated circulator or ferrite isolator in the feed line to the transmitter and duplexer
C. By installing a band-pass filter in the antenna feed line
D. By installing a low-pass filter in the antenna feed line

4AD-4.4 What can occur when a non-linear amplifier is used with an emission J3E transmitter?
A. Reduced amplifier efficiency
B. Increased intelligibility
C. Sideband inversion
D. Distortion

4AD-4.5 How can even-order harmonics be reduced or prevented in transmitter amplifier design?
A. By using a push-push amplifier
B. By using a push-pull amplifier
C. By operating class C
D. By operating class AB

4AD-5.1 What is <u>receiver</u> <u>desensitizing</u>?
A. A burst of noise when the squelch is set too low
B. A burst of noise when the squelch is set too high
C. A reduction in receiver sensitivity because of a strong signal on a nearby frequency
D. A reduction in receiver sensitivity when the AF gain control is turned down

4AD-5.2 What is the term used to refer to the reduction of receiver gain caused by the signals of a nearby station transmitting in the same frequency band?
A. Desensitizing
B. Quieting
C. Cross modulation interference
D. Squelch gain rollback

4AD-5.3   What is the term used to refer to a reduction in receiver sensitivity
           caused by unwanted high-level adjacent channel signals?
           A. Intermodulation distortion
           B. Quieting
           C. Desensitizing
           D. Overloading

4AD-5.4   What causes receiver desensitizing?
           A. Audio gain adjusted too low
           B. Squelch gain adjusted too high
           C. The presence of a strong signal on a nearby frequency
           D. Squelch gain adjusted too low

4AD-5.5   How can receiver desensitizing be reduced?
           A. Ensure good RF shielding between the transmitter and receiver
           B. Increase the transmitter audio gain
           C. Decrease the receiver squelch gain
           D. Increase the receiver bandwidth

4AD-6.1   What is cross-modulation interference?
           A. Interference between two transmitters of different modulation
              type
           B. Interference caused by audio rectification in the receiver preamp
           C. Harmonic distortion of the transmitted signal
           D. Modulation from an unwanted signal is heard in addition to the
              desired signal

4AD-6.2   What is the term used to refer to the condition where the signals from
           a very strong station are superimposed on other signals being
           received?
           A. Intermodulation distortion
           B. Cross-modulation interference
           C. Receiver quieting
           D. Capture effect

4AD-6.3   How can cross-modulation in a receiver be reduced?
           A. By installing a filter at the receiver
           B. By using a better antenna
           C. By increasing the receiver's RF gain while decreasing the AF
              gain
           D. By adjusting the pass-band tuning

4AD-6.4   What is the result of cross-modulation?
           A. A decrease in modulation level of transmitted signals
           B. Receiver quieting
           C. The modulation of an unwanted signal is heard on the desired
              signal
           D. Inverted sidebands in the final stage of the amplifier

4AD-7.1   What is the capture effect?
           A. All signals on a frequency are demodulated by an FM receiver
           B. All signals on a frequency are demodulated by an AM receiver
           C. The loudest signal received is the only demodulated signal
           D. The weakest signal received is the only demodulated signal

4AD-7.2   What is the term used to refer to the reception blockage of one parti-
           cular emission F3E signal by another emission F3E signal?
           A. Desensitization
           B. Cross-modulation interference
           C. Capture effect
           D. Frequency discrimination

4AD-7.3 With which emission type is the capture-effect most pronounced?
   A. FM
   B. SSB
   C. AM
   D. CW

4AE-1.1   What is reactive power?
           A. Wattless, non-productive power
           B. Power consumed in wire resistance in an inductor
           C. Power lost because of capacitor leakage
           D. Power consumed in circuit Q

4AE-1.2   What is the term for an out-of-phase, non-productive power assoc-
           iated with inductors and capacitors?
           A. Effective power
           B. True power
           C. Peak envelope power
           D. Reactive power

4AE-1.3   What is the term for energy that is stored in an electromagnetic or
           electrostatic field?
           A. Potential energy
           B. Amperes-joules
           C. Joules-coulombs
           D. Kinetic energy

4AE-1.4   What is responsible for the phenomenon when voltages across
           reactances in series can often be larger than the voltages applied
           to them?
           A. Capacitance
           B. Resonance
           C. Conductance
           D. Resistance

4AE-2.1   What is resonance in an electrical circuit?
           A. The highest frequency that will pass current
           B. The lowest frequency that will pass current
           C. The frequency at which capacitive reactance equals induc-
              tive reactance
           D. The frequency at which power factor is at a minimum

4AE-2.2   Under what conditions does resonance occur in an electrical
           circuit?
           A. When the power factor is at a minimum
           B. When inductive and capacitive reactances are equal
           C. When the square root of the sum of the capacitive and in-
              ductive reactances is equal to the resonant frequency
           D. When the square root of the product of the capacitive and
              inductive reactances is equal to the resonant frequency

4AE-2.3   What is the term for the phenomena which occurs in an electrical
           circuit when the inductive reactance equals the capacitive
           reactance?
           A. Reactive quiescence
           B. High Q
           C. Reactive equilibrium
           D. Resonance

4AE-2.4   What is the approximate magnitude of the impedance of a series
           R-L-C circuit at resonance?
           A. High, as compared to the circuit resistance
           B. Approximately equal to the circuit resistance
           C. Approximately equal to X_L
           D. Approximately equal to X_C

4AE-2.5 What is the approximate magnitude of the impedance of a parallel R-L-C circuit at resonance?
A. High, as compared to the circuit resistance
B. Approximately equal to X_L
C. Low, as compared to the circuit resistance
D. Approximately equal to X_C

4AE-2.6 What is the characteristic of the current flow in a series R-L-C circuit at resonance?
A. It is at a minimum
B. It is at a maximum
C. It is dc
D. It is zero

4AE-2.7 What is the characteristic of the current flow in a parallel R-L-C circuit at resonance?
A. The current circulating in the parallel elements is at a minimum
B. The current circulating in the parallel elements is at a maximum
C. The current circulating in the parallel elements is dc
D. The current circulating in the parallel elements is zero

4AE-3.1 What is the skin effect?
A. The phenomenon where RF current flows in a thinner layer of the conductor, close to the surface, as frequency increases
B. The phenomenon where RF current flows in a thinner layer of the conductor, close to the surface, as frequency decreases
C. The phenomenon where thermal effects on the surface of the conductor increase the impedance
D. The phenomenon where thermal effects on the surface of the conductor decrease the impedance

4AE-3.2 What is the term for the phenomenon where most of an rf current flows along the surface of the conductor?
A. Layer effect
B. Seeburg Effect
C. Skin effect
D. Resonance

*4AE-3.3 Where does practically all of rf current flow in a conductor?
A. Along the surface
B. In the center of the conductor
C. In the magnetic field around the conductor
D. In the electromagnetic field in the conductor center

4AE-3.4 Why does practically all of an rf current flow within a few thousandths-of-an-inch of the conductor's surface?
A. Because of skin effect
B. Because the RF resistance of the conductor is much less than the dc resistance
C. Because of heating of the metal at the conductor's interior
D. Because of the ac-resistance of the conductor's self inductance

4AE-3.5   Why is the resistance of a conductor different for rf current than
           for dc?
           A. Because the insulation conducts current at radio frequencies
           B. Because of the Heisenburg Effect
           C. Because of skin effect
           D. Because conductors are non-linear devices

4AE-4.1   What is a magnetic field?
           A. Current flow through space around a permanent magnet
           B. A force set up when current flows through a conductor
           C. The force between the plates of a charged capacitor
           D. The force that drives current through a resistor

4AE-4.2   In what direction is the magnetic field about a conductor when
           current is flowing?
           A. In the same direction as the current
           B. In a direction opposite to the current flow
           C. In all directions; omnidirectional
           D. In a direction determined by the left hand rule

4AE-4.3   What device is used to store electrical energy in an electrostatic
           field?
           A. A battery
           B. A transformer
           C. A capacitor
           D. An inductor

4AE-4.4   What is the term used to express the amount of electrical energy
           stored in an electrostatic field?
           A. Coulombs
           B. Joules
           C. Watts
           D. Volts

4AE-4.5   What factors determine the capacitance of a capacitor?
           A. Area of the plates, voltage on the plates and distance
              between the plates
           B. Area of the plates, distance between the plates and the die
              lectric constant of the material between the plates
           C. Area of the plates, voltage on the plates and the dielectric
              constant of the material between the plates
           D. Area of the plates, amount of charge on the plates and the
              dielectric constant of the material between the plates

*4AE-4.6   What is the dialectric constant for air?
           A. Approximately 1
           B. Approximately 2
           C. Approximately 4
           D. Approximately 0

4AE-4.7   What determines the strength of the magnetic field around a
           conductor?
           A. The resistance divided by the current
           B. The ratio of the current to the resistance
           C. The diameter of the conductor
           D. The amount of current

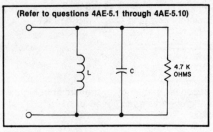

(Refer to questions 4AE-5.1 through 4AE-5.10)

4.7 K OHMS

FIGURE 4AE-5-1

*4AE-5.1 What is the resonant frequency of the circuit in Figure 4E-5-1 when L is 50 microhenrys and C is 40 picofarads?
A. 79.6 MHz
B. 1.78 MHz
C. 3.56 MHz
D. 7.96 MHz

*4AE-5.2 What is the resonant frequency of the circuit in Figure 4E-5-1 when L is 40 microhenrys and C is 200 picofarads?
A. 1.99 kHz
B. 1.78 MHz
C. 1.99 MHz
D. 1.78 kHz

*4AE-5.3 What is the resonant frequency of the circuit in Figure 4E-5-1 when L is 50 microhenrys and C is 10 picofarads?
A. 3.18 MHz
B. 3.18 kHz
C. 7.12 MHz
D. 7.12 kHz

*4AE-5.4 What is the resonant frequency of the circuit in Figure 4E-5-1 when L is 25 microhenrys and C is 10 picofarads?
A. 10.1 MHz
B. 63.7 MHz
C. 10.1 kHz
D. 63.7 kHz

*4AE-5.5 What is the resonant frequency of the circuit in Figure 4E-5-1 when L is 3 microhenrys and C is 40 picofarads?
A. 13.1 MHz
B. 14.5 MHz
C. 14.5 kHz
D. 13.1 kHz

**Figure 4E-5-1 when L is 4 microhenrys and C is 20 picofarads?
A. 19.9 kHz
B. 17.8 kHz
C. 19.9 MHz
D. 17.8 MHz

** This question is missing the question number and part of the question. We believe the question should read: "4AE-5.6 What is the resonant frequency of the circuit in Figure 4AE-5-1 when L is 4 microhenrys and C is 20 picofarads?"

*4AE-5.7   What is the resonant frequency of the circuit in Figure 4E-5-1
when L is 8 microhenrys and C is 7 picofarads?
 A. 2.84 MHz
 B. 28.4 MHz
 C. 21.3 MHz
 D. 2.13 MHz

*4AE-5.8   What is the resonant frequency of the circuit in Figure 4E-5-1
when L is 3 microhenrys and C is 15 picofarads?
 A. 23.7 MHz
 B. 23.7 kHz
 C. 35.4 kHz
 D. 35.4 MHz

*4AE-5.9   What is the resonant frequency of the circuit in Figure 4E-5-1
when L is 4 microhenrys and C is 8 picofarads?
 A. 28.1 kHz
 B. 28.1 MHz
 C. 49.7 MHz
 D. 49.7 kHz

*4AE-5.10   What is the resonant frequency of the circuit in Figure 4E-5-1
when L is 1 microhenry and C is 9 picofarads?
 A. 17.7 MHz
 B. 17.7 kHz
 C. 53.1 MHz
 D. 53.1 kHz

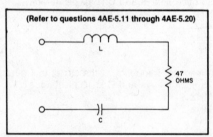

(Refer to questions 4AE-5.11 through 4AE-5.20)

FIGURE 4AE-5-2

4AE-5.11   What is the resonant frequency of the circuit in Figure 4AE-5-2
when L is 1 microhenry and C is 10 picofarads?
 A. 50.3 MHz
 B. 15.9 MHz
 C. 15.9 kHz
 D. 50.3 kHz

4AE-5.12   What is the resonant frequency of the circuit in Figure 4AE-5-2
when L is 2 microhenrys and C is 15 picofarads?
 A. 29.1 kHz
 B. 29.1 MHz
 C. 5.31 MHz
 D. 5.31 kHz

4AE-5.13   What is the resonant frequency of the circuit in Figure 4AE-5-2 when L is 5 microhenrys and C is 9 picofarads?
A. 23.7 kHz
B. 3.54 kHz
C. 23.7 MHz
D. 3.54 MHz

4AE-5.14   What is the resonant frequency of the circuit in Figure 4AE-5-2 when L is 2 microhenrys and C is 30 picofarads?
A. 2.65 kHz
B. 20.5 kHz
C. 2.65 MHz
D. 20.5 MHz

4AE-5.15   What is the resonant frequency of the circuit in Figure 4AE-5-2 when L is 15 microhenrys and C is 5 picofarads?
A. 18.4 MHz
B. 2.12 MHz
C. 18.4 kHz
D. 2.12 kHz

4AE-5.16   What is the resonant frequency of the circuit in Figure 4AE-5-2 when L is 3 microhenrys and C is 40 picofarads?
A. 1.33 kHz
B. 14.5 MHz
C. 1.33 MHz
D. 14.5 kHz

4AE-5.17   What is the resonant frequency of the circuit in Figure 4AE-5-2 when L is 40 microhenrys and C is 6 picofarads?
A. 6.63 MHz
B. 6.63 kHz
C. 10.3 MHz
D. 10.3 kHz

4AE-5.18   What is the resonant frequency of the circuit in Figure 4AE-5-2 when L is 10 microhenrys and C is 50 picofarads?
A. 3.18 MHz
B. 3.18 kHz
C. 7.12 kHz
D. 7.12 MHz

4AE-5.19   What is the resonant frequency of the circuit in Figure 4AE-5-2 when L is 200 microhenrys and C is 10 picofarads?
A. 3.56 MHz
B. 7.96 kHz
C. 3.56 kHz
D. 7.96 MHz

4AE-5.20   What is the resonant frequency of the circuit in Figure 4AE-5-2 when L is 90 microhenrys and C is 100 picofarads?
A. 1.77 MHz
B. 1.68 MHz
C. 1.77 kHz
D. 1.68 kHz

4AE-5.21   What is the half-power bandwidth of a parallel resonant circuit
            which has a resonant frequency of 1.8 MHz and a Q of 95?
            A. 18.9 kHz
            B. 1.89 kHz
            C. 189 Hz
            D. 58.7 kHz

4AE-5.22   What is the half-power bandwidth of a parallel resonant circuit
            which has a resonant frequency of 3.6 MHz and a Q of 218?
            A. 58.7 kHz
            B. 606 kHz
            C. 47.3 kHz
            D. 16.5 kHz

4AE-5.23   What is the half-power bandwidth of a parallel resonant circuit
            which has a resonant frequency of 7.1 MHz and a Q of 150?
            A. 211 kHz
            B. 16.5 kHz
            C. 47.3 kHz
            D. 21.1 kHz

4AE-5.24   What is the half-power bandwidth of a parallel resonant circuit
            which has a resonant frequency of 12.8 MHz and a Q of 218?
            A. 21.1 kHz
            B. 27.9 kHz
            C. 17 kHz
            D. 58.7 kHz

4AE-5.25   What is the half-power bandwidth of a parallel resonant circuit
            which has a resonant frequency of 14.25 MHz and a Q of 150?
            A. 95 kHz
            B. 10.5 kHz
            C. 10.5 MHz
            D. 17 kHz

4AE-5.26   What is the half-power bandwidth of a parallel resonant circuit
            which has a resonant frequency of 21.15 MHz and a Q of 95?
            A. 4.49 kHz
            B. 44.9 kHz
            C. 22.3 kHz
            D. 222.6 kHz

4AE-5.27   What is the half-power bandwidth of a parallel resonant circuit
            which has a resonant frequency of 10.1 MHz and a Q of 225?
            A. 4.49 kHz
            B. 44.9 kHz
            C. 22.3 kHz
            D. 223 kHz

4AE-5.28   What is the half-power bandwidth of a parallel resonant circuit
            which has a resonant frequency of 18.1 MHz and a Q of 195?
            A. 92.8 kHz
            B. 10.8 kHz
            C. 22.3 kHz
            D. 44.9 kHz

4AE-5.29 What is the half-power bandwidth of a parallel resonant circuit
which has a resonant frequency of 3.7 MHz and a Q of 118?
    A. 22.3 kHz
    B. 76.2 kHz
    C. 31.4 kHz
    D. 10.8 kHz

4AE-5.30 What is the half-power bandwidth of a parallel resonant circuit
which has a resonant frequency of 14.25 MHz and a Q of 187?
    A. 22.3 kHz
    B. 10.8 kHz
    C. 13.1 kHz
    D. 76.2 kHz

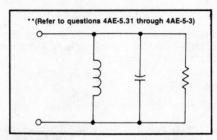

**(Refer to questions 4AE-5.31 through 4AE-5-3)

FIGURE 4AE-5-3

**This Figure actually refers to questions 4AE-5.31 through 4AE-5.40.

4AE-5.31 What is the Q of the circuit in Figure 4AE-5-3 when the resonant
frequency is 14.128 MHz, the inductance is 2.7 microhenrys and
the resistance is 18,000 ohms?
    A. 75.1
    B. 7.51
    C. 71.5
    D. 0.013

4AE-5.32 What is the Q of the circuit in Figure 4AE-5-3 when the resonant
frequency is 14.128 MHz, the inductance is 4.7 microhenrys and
the resistance is 18,000 ohms?
    A. 4.31
    B. 43.1
    C. 13.3
    D. 0.023

4AE-5.33 What is the Q of the circuit in Figure 4AE-5-3 when the resonant
frequency is 4.468 MHz, the inductance is 47 microhenrys and the
resistance is 180 ohms?
    A. 0.00735
    B. 7.35
    C. 0.136
    D. 13.3

4AE-5.34 What is the Q of the circuit in Figure 4AE-5-3 when the resonant
frequency is 14.225 MHz, the inductance is 3.5 microhenrys and
the resistance is 10,000 ohms?
    A. 7.35
    B. 0.0319
    C. 71.5
    D. 31.9

4AE-5.35 What is the Q of the circuit in Figure 4AE-5-3 when the resonant frequency is 7.125 MHz, the inductance is 8.2 microhenrys and the resistance is 1,000 ohms?
  A. 36.8
  B. 0.273
  C. 0.368
  D. 2.73

4AE-5.36 What is the Q of the circuit in Figure 4AE-5-3 when the resonant frequency is 7.125 MHz, the inductance is 10.1 microhenrys and the resistance is 100 ohms?
  A. 0.221
  B. 4.52
  C. 0.00452
  D. 22.1

4AE-5.37 What is the Q of the circuit in Figure 4AE-5-3 when the resonant frequency is 7.125 MHz, the inductance is 12.6 microhenrys and the resistance is 22,000 ohms?
  A. 22.1
  B. 39
  C. 25.6
  D. 0.0256

4AE-5.38 What is the Q of the circuit in Figure 4AE-5-3 when the resonant frequency is 3.625 MHz, the inductance is 3 microhenrys and the resistance is 2,200 ohms?
  A. 0.031
  B. 32.2
  C. 31.1
  D. 25.6

4AE-5.39 What is the Q of the circuit in Figure 4AE-5-3 when the resonant frequency is 3.625 MHz, the inductance is 42 microhenrys and the resistance is 220 ohms?
  A. 23
  B. 0.00435
  C. 4.35
  D. 0.23

4AE-5.40 What is the Q of the circuit in Figure 4AE-5-3 when the resonant frequency is 3.625 MHz, the inductance is 43 microhenrys and the resistance is 1,800 ohms?
  A. 1.84
  B. 0.543
  C. 54.3
  D. 23

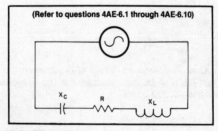

FIGURE 4AE-6

4AE-6.1   What is the phase angle between the voltage across and the
           current through the circuit in Figure 4AE-6, when Xc is 25 ohms,
           R is 100 ohms, and Xl is 100 ohms?
           A. 36.9 degrees with the voltage leading the current
           B. 53.1 degrees with the voltage lagging the current
           C. 36.9 degrees with the voltage lagging the current
           D. 53.1 degrees with the voltage leading the current

4AE-6.2   What is the phase angle between the voltage across and the
           current through the circuit in Figure 4AE-6, when Xc is 25 ohms,
           R is 100 ohms, and Xl is 50 ohms?
           A. 14 degrees with the voltage lagging the current
           B. 14 degrees with the voltage leading the current
           C. 76 degrees with the voltage lagging the current
           D. 76 degrees with the voltage leading the current

4AE-6.3   What is the phase angle between the voltage across and the
           current through the circuit in Figure 4AE-6, when Xc is 500 ohms,
           R is 1000 ohms, and Xl is 250 ohms?
           A. 68.2 degrees with the voltage leading the current
           B. 14.1 degrees with the voltage leading the current
           C. 14.1 degrees with the voltage lagging the current
           D. 68.2 degrees with the voltage lagging the current

4AE-6.4   What is the phase angle between the voltage across and the
           current through the circuit in Figure 4AE-6, when Xc is 75 ohms,
           R is 100 ohms, and Xl is 100 ohms?
           A. 76 degrees with the voltage leading the current
           B. 14 degrees with the voltage leading the current
           C. 14 degrees with the voltage lagging the current
           D. 76 degrees with the voltage lagging the current

4AE-6.5   What is the phase angle between the voltage across and the
           current through the circuit in Figure 4AE-6, when Xc is 50 ohms,
           R is 100 ohms, and Xl is 25 ohms?
           A. 76 degrees with the voltage lagging the current
           B. 14 degrees with the voltage leading the current
           C. 76 degrees with the voltage leading the current
           D. 14 degrees with the voltage lagging the current

4AE-6.6   What is the phase angle between the voltage across and the
           current through the circuit in Figure 4AE-6, when Xc is 75 ohms,
           R is 100 ohms, and Xl is 50 ohms?
           A. 76 degrees with the voltage lagging the current
           B. 14 degrees with the voltage lagging the current
           C. 14 degrees with the voltage leading the current
           D. 76 degrees with the voltage leading the current

4AE-6.7   What is the phase angle between the voltage across and the
           current through the circuit in Figure 4AE-6, when Xc is 100 ohms,
           R is 100 ohms, and Xl is 75 ohms?
           A. 14 degrees with the voltage lagging the current
           B. 14 degrees with the voltage leading the current
           C. 76 degrees with the voltage leading the current
           D. 76 degrees with the voltage lagging the current

4AE-6.8　What is the phase angle between the voltage across and the current through the circuit in Figure 4AE-6, when Xc is 250 ohms, R is 1000 ohms, and Xl is 500 ohms?
A. 81.47 degrees with the voltage lagging the current
B. 81.47 degrees with the voltage leading the current
C. 14.04 degrees with the voltage lagging the current
D. 14.04 degrees with the voltage leading the current

4AE-6.9　What is the phase angle between the voltage across and the current through the circuit in Figure 4AE-6, when Xc is 50 ohms, R is 100 ohms, and Xl is 75 ohms?
A. 76 degrees with the voltage leading the current
B. 76 degrees with the voltage lagging the current
C. 14 degrees with the voltage lagging the current
D. 14 degrees with the voltage leading the current

4AE-6.10　What is the phase angle between the voltage across and the current through the circuit in Figure 4AE-6, when Xc is 100 ohms, R is 100 ohms, and Xl is 25 ohms?
A. 36.9 degrees with the voltage leading the current
B. 53.1 degrees with the voltage lagging the current
C. 36.9 degrees with the voltage lagging the current
D. 53.1 degrees with the voltage leading the current

4AE-7.1　Why would the rate at which electrical energy is used in a circuit be less than the product of the magnitudes of the ac voltage and current?
A. Because there is a phase angle that is greater than zero between the current and voltage
B. Because there are only resistances in the circuit
C. Because there are no reactances in the circuit
D. Because there is a phase angle that is equal to zero between the current and voltage

4AE-7.2　In a circuit where the ac voltage and current are out of phase, how can the true power be determined?
A. By multiplying the apparent power times the power factor
B. By subtracting the apparent power from the power factor
C. By dividing the apparent power by the power factor
D. By multiplying the RMS voltage times the RMS current

4AE-7.3　What does the power factor equal in an R-L circuit having a 60 degree phase angle between the voltage and the current?
A. 1.414
B. 0.866
C. 0.5
D. 1.73

4AE-7.4　What does the power factor equal in an R-L circuit having a 45 degree phase angle between the voltage and the current?
A. 0.866
B. 1.0
C. 0.5
D. 0.707

4AE-7.5　What does the power factor equal in an R-L circuit having a 30 degree phase angle between the voltage and the current?
A. 1.73
B. 0.5
C. 0.866
D. 0.577

4AE-7.6 How many watts are being consumed in a circuit having a power factor of 0.2 when the input is 100-vac and 4-amperes is being drawn?
A. 400 watts
B. 80 watts
C. 2000 watts
D. 50 watts

4AE-7.7 How many watts are being consumed in a circuit having a power factor of 0.6 when the input is 200-vac and 5-amperes is being drawn?
A. 200 watts
B. 1000 watts
C. 1600 watts
D. 600 watts

4AE-8.1 What is the effective radiated power of a station in repeater operation with 50 watts transmitter power output, 4 dB feedline loss, 3 dB duplexer and circulator loss, and 6 dB antenna gain?
A. 158 watts, assuming the antenna gain is referenced to a half-wave dipole
B. 39.7 watts, assuming the antenna gain is referenced to a half-wave dipole
C. 251 watts, assuming the antenna gain is referenced to a half-wave dipole
D. 69.9 watts, assuming the antenna gain is referenced to a half-wave dipole

4AE-8.2 What is the effective radiated power of a station in repeater operation with 50 watts transmitter power output, 5 dB feedline loss, 4 dB duplexer and circulator loss, and 7 dB antenna gain?
A. 300 watts, assuming the antenna gain is referenced to a half-wave dipole
B. 315 watts, assuming the antenna gain is referenced to a half-wave dipole
C. 31.5 watts, assuming the antenna gain is referenced to a half-wave dipole
D. 69.9 watts, assuming the antenna gain is referenced to a half-wave dipole

4AE-8.3 What is the effective radiated power of a station in repeater operation with 75 watts transmitter power output, 4 dB feedline loss, 3 dB duplexer and circulator loss, and 10 dB antenna gain?
A. 600 watts, assuming the antenna gain is referenced to a half-wave dipole
B. 75 watts, assuming the antenna gain is referenced to a half-wave dipole
C. 18.75 watts, assuming the antenna gain is referenced to a half-wave dipole
D. 150 watts, assuming the antenna gain is referenced to a half-wave dipole

*4AE-8.4 What is the effective radiated power of a station in repeater operation with 75 watts transmitter power output, 5 dB feedline loss, 4 dB duplexer and circulator loss, and 6 dB antenna gain.?
  A. 37.6 watts, assuming the antenna gain is referenced to a half-wave dipole
  B. 237 watts, assuming the antenna gain is referenced to a half-wave dipole
  C. 150 watts, assuming the antenna gain is referenced to a half-wave dipole
  D. 23.7 watts, assuming the antenna gain is referenced to a half-wave dipole

4AE-8.5 What is the effective radiated power of a station in repeater operation with 100 watts transmitter power output, 4 dB feedline loss, 3 dB duplexer and circulator loss, and 7 dB antenna gain?
  A. 631 watts, assuming the antenna gain is referenced to a half-wave dipole
  B. 400 watts, assuming the antenna gain is referenced to a half-wave dipole
  C. 25 watts, assuming the antenna gain is referenced to a half-wave dipole
  D. 100 watts, assuming the antenna gain is referenced to a half-wave dipole

4AE-8.6 What is the effective radiated power of a station in repeater operation with 100 watts transmitter power output, 5 dB feedline loss, 4 dB duplexer and circulator loss, and 10 dB antenna gain?
  A. 800 watts, assuming the antenna gain is referenced to a half-wave dipole
  B. 126 watts, assuming the antenna gain is referenced to a half-wave dipole
  C. 12.5 watts, assuming the antenna gain is referenced to a half-wave dipole
  D. 1260 watts, assuming the antenna gain is referenced to a half-wave dipole

4AE-8.7 What is the effective radiated power of a station in repeater operation with I20 watts transmitter power output, 5 dB feedline loss, 4 dB duplexer and circulator loss, and 6 dB antenna gain?
  A. 60l watts, assuming the antenna gain is referenced to a half-wave dipole
  B. 240 watts, assuming the antenna gain is referenced to a half-wave dipole
  C. 60 watts, assuming the antenna gain is referenced to a half-wave dipole
  D. 379 watts, assuming the antenna gain is referenced to a half-wave dipole

4AE-8.8 What is the effective radiated power of a station in repeater operation with 150 watts transmitter power output, 4 dB feedline loss, 3 dB duplexer and circulator loss, and 7 dB antenna gain?
  A. 946 watts, assuming the antenna gain is referenced to a half-wave dipole
  B. 37.5 watts, assuming the antenna gain is referenced to a half-wave dipole
  C. 600 watts, assuming the antenna gain is referenced to a half-wave dipole
  D. 150 watts, assuming the antenna gain is referenced to a half-wave dipole

4AE-8.9   What is the effective radiated power of a station in repeater operation with 200 watts transmitter power output, 4 dB feedline loss, 4 dB duplexer and circulator loss, and 10 dB antenna gain?
   A. 317 watts, assuming the antenna gain is referenced to a half-wave dipole
   B. 2000 watts, assuming the antenna gain is referenced to a half-wave dipole
   C. 126 watts, assuming the antenna gain is referenced to a half-wave dipole
   D. 260 watts, assuming the antenna gain is referenced to a half-wave dipole

4AE-8.10   What is the effective radiated power of a station in repeater operation with 200 watts transmitter power output, 4 dB feedline loss, 3 dB duplexer and circulator loss, and 6 dB antenna gain?
   A. 252 watts, assuming the antenna gain is referenced to a half-wave dipole
   B. 63.2 watts, assuming the antenna gain is referenced to a half-wave dipole
   C. 632 watts, assuming the antenna gain is referenced to a half-wave dipole
   D. 159 watts, assuming the antenna gain is referenced to a half-wave dipole

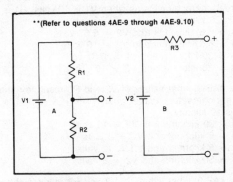

**(Refer to questions 4AE-9 through 4AE-9.10)

FIGURE 4AE-9

**This Figure actually refers to questions 4AE-9.1 through 4AE-9.10.

4AE-9.1   In Figure 4AE-9, what values of V2 and R3 result in the same voltage and current characteristics as when V1 is 8-volts, R1 is 8 kilohms, and R2 is 8 kilohms?
   A. R3 = 4 kilohms and V2 = 8 volts
   B. R3 = 4 kilohms and V2 = 4 volts
   C. R3 = 16 kilohms and V2 = 8 volts
   D. R3 = 16 kilohms and V2 = 4 volts

4AE-9.2   In Figure 4AE-9, what values of V2 and R3 result in the same voltage and current characteristics as when V1 is 8-volts, R1 is 16 kilohms, and R2 is 8 kilohms?
   A. R3 = 24 kilohms and V2 = 5.33 volts
   B. R3 = 5.33 kilohms and V2 = 8 volts
   C. R3 = 5.33 kilohms and V2 = 2.67 volts
   D. R3 = 24 kilohms and V2 = 8 volts

4AE-9.3   In Figure 4AE-9, what values of V2 and R3 result in the same voltage
           and current characteristics as when V1 is 8-volts, R1 is 8 kilohms, and
           R2 is 16 kilohms?
           A. R3 = 24 kilohms and V2 = 8 volts
           B. R3 = 8 kilohms and V2 = 4 volts
           C. R3 = 5.33 kilohms and V2 = 5.33 volts
           D. R3 = 5.33 kilohms and V2 = 8 volts

4AE-9.4   In Figure 4AE-9, what values of V2 and R3 result in the same voltage
           and current characteristics as when V1 is 10-volts, R1 is 10 kilohms,
           and R2 is 10 kilohms?
           A. R3 = 10 kilohms and V2 = 5 volts
           B. R3 = 20 kilohms and V2 = 5 volts
           C. R3 = 20 kilohms and V2 = 10 volts
           D. R3 = 5 kilohms and V2 = 5 volts

4AE-9.5   In Figure 4AE-9, what values of V2 and R3 result in the same voltage
           and current characteristics as when V1 is 10-volts, R1 is 20 kilohms,
           and R2 is 10 kilohms?
           A. R3 = 30 kilohms and V2 = 10 volts
           B. R3 = 6.67 kilohms and V2 = 10 volts
           C. R3 = 6.67 kilohms and V2 = 3.33 volts
           D. R3 = 30 kilohms and V2 = 3.33 volts

4AE-9.6   In Figure 4AE-9, what values of V2 and R3 result in the same voltage
           and current characteristics as when V1 is 10-volts, R1 is 10 kilohms,
           and R2 is 20 kilohms?
           A. R3 = 6.67 kilohms and V2 = 6.67 volts
           B. R3 = 6.67 kilohms and V2 = 10 volts
           C. R3 = 30 kilohms and V2 = 6.67 volts
           D. R3 = 30 kilohms and V2 = 10 volts

4AE-9.7   In Figure 4AE-9, what values of V2 and R3 result in the same voltage
           and current characteristics as when V1 is 12-volts, R1 is 10 kilohms,
           and R2 is 10 kilohms?
           A. R3 = 20 kilohms and V2 = 12 volts
           B. R3 = 5 kilohms and V2 = 6 volts
           C. R3 = 5 kilohms and V2 = 12 volts
           D. R3 = 30 kilohms and V2 = 6 volts

4AE-9.8   In Figure 4AE-9, what values of V2 and R3 result in the same voltage
           and current characteristics as when V1 is 12-volts, R1 is 20 kilohms,
           and R2 is 10 kilohms?
           A. R3 = 30 kilohms and V2 = 4 volts
           B. R3 = 6.67 kilohms and V2 = 4 volts
           C. R3 = 30 kilohms and V2 = 12 volts
           D. R3 = 6.67 kilohms and V2 = 12 volts

4AE-9.9   In Figure 4AE-9, what values of V2 and R3 result in the same voltage
           and current characteristics as when V1 is 12-volts, R1 is 10 kilohms,
           and R2 is 20 kilohms?
           A. R3 = 6.67 kilohms and V2 = 12 volts
           B. R3 = 30 kilohms and V2 = 12 volts
           C. R3 = 6.67 kilohms and V2 = 8 volts
           D. R3 = 30 kilohms and V2 = 8 volts

4AE-9.10   In Figure 4AE-9, what values of V2 and R3 result in the same voltage
           and current characteristics as when V1 is 12-volts, R1 is 20 kilohms,
           and R2 is 20 kilohms?
           A. R3 = 40 kilohms and V2 = 12 volts
           B. R3 = 40 kilohms and V2 = 6 volts
           C. R3 = 10 kilohms and V2 = 6 volts
           D. R3 = 10 kilohms and V2 = 12 volts

SUBELEMENT 4AF — Circuit Components (6 questions)

4AF-1.1 What is the schematic symbol for a semiconductor diode/rectifier?

A.   B.

C. ⊣⊢   D.

4AF-1.2 Structurally, what are the two main categories of semiconductor diodes?
- A. Junction and point contact
- B. Electrolytic and junction
- C. Electrolytic and point contact
- D. Vacuum and point contact

4AF-1.3 What is the schematic symbol for a Zener diode?

A.   B.

C.   D.

4AF-1.4 What are the two primary classifications of Zener diodes?
- A. Hot carrier and tunnel
- B. Varactor and rectifying
- C. Voltage regulator and voltage reference
- D. Forward and reversed biased

4AF-1.5 What is the principal characteristic of a Zener diode?
- A. A constant current under conditions of varying voltage
- B. A constant voltage under conditions of varying current
- C. A negative resistance region
- D. An internal capacitance that varies with the applied voltage

4AF-1.6 What is the range of voltage ratings available in Zener diodes?
- A. 2.4 volts to 200 volts
- B. 1.2 volts to 7 volts
- C. 3 volts to 2000 volts
- D. 1.2 volts to 5.6 volts

4AF-1.7 What is the schematic symbol for a tunnel diode?

A.   B.

C.   D.

4AF-1.8   What is the principal characteristic of a tunnel diode?
           A. A high forward resistance
           B. A very high PIV
           C. A negative resistance region
           D. A high forward current rating

4AF-1.9   What special type of diode is capable of both amplification and
           oscillation?
           A. Point contact diodes
           B. Zener diodes
           C. Tunnel diodes
           D. Junction diodes

4AF-1.10   What is the schematic symbol for a varactor diode?

4AF-1.11   What type of semiconductor diode varies its internal capacitance
           as the voltage applied to its terminals varies?
           A. A varactor diode
           B. A tunnel diode
           C. A silicon-controlled rectifier
           D. A Zener diode

4AF-1.12   What is the principal characteristic of a varactor diode?
           A. It has a constant voltage under conditions of varying current
           B. Its internal capacitance varies with the applied voltage
           C. It has a negative resistance region
           D. It has a very high PIV

4AF-1.13   What is a common use of a varactor diode?
           A. As a constant current source
           B. As a constant voltage source
           C. As a voltage controlled inductance
           D. As a voltage controlled capacitance

4AF-1.14   What is a common use of a hot-carrier diode?
           A. As balanced mixers in SSB generation
           B. As a variable capacitance in an automatic frequency control
              circuit
           C. As a constant voltage reference in a power supply
           D. As VHF and UHF mixers and detectors

4AF-1.15   What limits the maximum forward current in a junction diode?
           A. The peak inverse voltage
           B. The junction temperature
           C. The forward voltage
           D. The back EMF

4AF-1.16 How are junction diodes rated?
    A. Maximum forward current and capacitance
    B. Maximum reverse current and PIV
    C. Maximum reverse current and capacitance
    D. Maximum forward current and PIV

4AF-1.17 What is a common use for point contact diodes?
    A. As a constant current source
    B. As a constant voltage source
    C. As an RF detector
    D. As a high voltage rectifier

4AF-1.18 What type of diode is made of a metal whisker touching a very small semi-conductor die?
    A. Zener diode
    B. Varactor diode
    C. Junction diode
    D. Point contact diode

*4AF-1.19 What is common use for PIN diodes?
    A. As a constant current source
    B. As a constant voltage source
    C. As an RF switch
    D. As a high voltage rectifier

*4AF-1.20 What special type of diode is often use for RF switches, attenuators, and various types of phase shifting devices?
    A. Tunnel diodes
    B. Varactor diodes
    C. PIN diodes
    D. Junction diodes

4AF-2.1 What is the schematic symbol for a PNP transistor?

 A.   B.

 C.   D.

4AF-2.2 What is the schematic symbol for an NPN transistor?

 A.   B.

 C.   D.

4AF-2.3 What are the three terminals of a bipolar transistor?
    A. Cathode, plate and grid
    B. Base, collector and emitter
    C. Gate, source and sink
    D. Input, output and ground

4AF-2.4   What is the meaning of the term <u>alpha</u> with regard to bipolar transistors?
          A. The change of collector current with respect to base current
          B. The change of base current with respect to collector current
          C. The change of collector current with respect to emitter current
          D. The change of collector current with respect to gate current

4AF-2.5   What is the term used to express the ratio of change in dc collector current to a change in emitter current in a bipolar transistor?
          A. Gamma
          B. Epsilon
          C. Alpha
          D. Beta

4AF-2.6   What is the meaning of the term <u>beta</u> with regard to bipolar transistors?
          A. The change of collector current with respect to base current
          B. The change of base current with respect to emitter current
          C. The change of collector current with respect to emitter current
          D. The change in base current with respect to gate current

4AF-2.7   What is the term used to express the ratio of change in the dc collector current to a change in base current in a bipolar transistor?
          A. Alpha
          B. Beta
          C. Gamma
          D. Delta

4AF-2.8   What is the meaning of the term <u>alpha</u> <u>cutoff</u> <u>frequency</u> with regard to bipolar transistors?
          A. The practical lower frequency limit of a transistor in common emitter configuration
          B. The practical upper frequency limit of a transistor in common base configuration
          C. The practical lower frequency limit of a transistor in common base configuration
          D. The practical upper frequency limit of a transistor in common emitter configuration

4AF-2.9   What is the term used to express that frequency at which the grounded base current gain has decreased to 0.7 of the gain obtainable at 1 kHz in a transistor?
          A. Corner frequency
          B. Alpha cutoff frequency
          C. Beta cutoff frequency
          D. Alpha rejection frequency

4AF-2.10   What is the meaning of the term <u>beta cutoff frequency</u> with regard to a bipolar transistor?
  A. That frequency at which the grounded base current gain has decreased to 0.7 of that obtainable at 1 kHz in a transistor
  B. That frequency at which the grounded emitter current gain has decreased to 0.7 of that obtainable at 1 kHz in a transistor
  C. That frequency at which the grounded collector current gain has decreased to 0.7 of that obtainable at 1 kHz in a transistor
  D. That frequency at which the grounded gate current gain has decreased to 0.7 of that obtainable at 1 kHz in a transistor

4AF-2.11   What is the meaning of the term <u>transition region</u> with regard to a transistor?
  A. An area of low charge density around the P-N junction
  B. The area of maximum P-type charge
  C. The area of maximum N-type charge
  D. The point where wire leads are connected to the P- or N-type material

4AF-2.12   What does it mean for a transistor to be <u>fully saturated</u>?
  A. The collector current is at its maximum value
  B. The collector current is at its minimum value
  C. The transistor's Alpha is at its maximum value
  D. The transistor's Beta is at its maximum value

4AF-2.13   What does it mean for a transistor to be <u>cut off</u>?
  A. There is no base current
  B. The transistor is at its operating point
  C. No current flows from emitter to collector
  D. Maximum current flows from emitter to collector

4AF-2.14   What is the schematic symbol for a unijunction transistor?

A.   B.

C.   D.

4AF-2.15   What are the elements of a unijunction transistor?
  A. Base 1, base 2 and emitter
  B. Gate, cathode and anode
  C. Gate, base 1 and base 2
  D. Gate, source and sink

4AF-2.16   For best efficiency and stability, where on the load-line should a solid-state power amplifier be operated?
  A. Just below the saturation point
  B. Just above the saturation point
  C. At the saturation point
  D. At 1.414 times the saturation point

*4AF-2.17 What two elements widely used in semiconductor devices exhibit
          both metalic and non-metalic characteristics?
          A. Silicon and gold
          B. Silicon and germanium
          C. Galena and germanium
          D. Galena and bismuth

4AF-3.1   What is the schematic symbol for a silicon controlled rectifier?

A.   B.

C.   D.

4AF-3.2   What are the three terminals of an SCR?
          A. Anode, cathode and gate
          B. Gate, source and sink
          C. Base, collector and emitter
          D. Gate, base 1 and base 2

4AF-3.3   What are the two stable operating conditions of an SCR?
          A. Conducting and nonconducting
          B. Oscillating and quiescent
          C. Forward conducting and reverse conducting
          D. NPN conduction and PNP conduction

4AF-3.4   When an SCR is in the triggered or on condition, its electrical
          characteristics are similar to what other solid-state device (as
          measured between its cathode and anode)?
          A. The junction diode
          B. The tunnel diode
          C. The hot-carrier diode
          D. The varactor diode

4AF-3.5   Under what operating condition does an SCR exhibit electrical
          characteristics similar to a foward-biased silicon rectifier?
          A. During a switching transition
          B. When it is used as a detector
          C. When it is gated "off"
          D. When it is gated "on"

4AF-3.6   What is the schematic symbol for a TRIAC?

A.   B.

C.   D.

4AF-3.7 What is the transistor called which is fabricated as two complementary SCRs in parallel with a common gate terminal?
A. TRIAC
B. Bilateral SCR
C. Unijunction transistor
D. Field effect transistor

4AF-3.8 What are the three terminals of a TRIAC?
A. Emitter, base 1 and base 2
B. Gate, anode 1 and anode 2
C. Base, emitter and collector
D. Gate, source and sink

4AF-4.1 What is the schematic symbol for a light-emitting diode?

4AF-4.2 What is the normal operating voltage and current for a light-emitting diode?
A. 60 volts and 20 mA
B. 5 volts and 50 mA
C. 1.7 volts and 20 mA
D. 0.7 volts and 60 mA

4AF-4.3 What type of bias is required for an LED to produce luminescence?
A. Reverse bias
B. Forward bias
C. Zero bias
D. Inductive bias

4AF-4.4 What are the advantages of using an LED?
A. Low power consumption and long life
B. High lumens per cm per cm and low power consumption
C. High lumens per cm per cm and low voltage requirement
D. A current flows when the device is exposed to a light source

4AF-4.5 What colors are available in LEDs?
A. Yellow, blue, red, brown and green
B. Red, violet, yellow, white and green
C. Violet, blue, yellow, orange and red
D. Red, green, orange, white and yellow

4AF-4.6 What is the schematic symbol for a neon lamp?

4AF-4.7   What type neon lamp is usually used in amateur radio work?
           A. NE-1
           B. NE-2
           C. NE-3
           D. NE-4

4AF-4.8   What is the dc starting voltage for an NE-2 neon lamp?
           A. Approximately 67 volts
           B. Approximately 5 volts
           C. Approximately 5.6 volts
           D. Approximately 110 volts

4AF-4.9   What is the ac starting voltage for an NE-2 neon lamp?
           A. Approximately 110-V ac RMS
           B. Approximately 5-V ac RMS
           C. Approximately 5.6-V ac RMS
           D. Approximately 48-V ac RMS

4AF-4.10   How can a neon lamp be used to check for the presence of rf?
           A. A neon lamp will go out in the presence of RF
           B. A neon lamp will change color in the presence of RF
           C. A neon lamp will light only in the presence of very low fre-
              quency RF
           D. A neon lamp will light in the presence of RF

4AF-5.1   What would be the bandwidth of a good crystal lattice band-pass
           filter for emission J3E?
           A. 6 kHz at −6 dB
           B. 2.1 kHz at −6 dB
           C. 500 Hz at −6 dB
           D. 15 kHz at −6 dB

4AF-5.2   What would be the bandwidth of a good crystal lattice band-pass filter
           for emission A3E?
           A. 1 kHz at −6 dB
           B. 500 Hz at −6 dB
           C. 6 kHz at −6 dB
           D. 15 kHz at −6 dB

4AF-5.3   What is a crystal lattice filter?
           A. A power supply filter made with crisscrossed quartz crystals
           B. An audio filter made with 4 quartz crystals at 1-kHz intervals
           C. A filter with infinitely wide and shallow skirts made using quartz
              crystals
           D. A filter with narrow bandwidth and steep skirts made using
              quartz crystals

4AF-5.4   What technique can be used to construct low cost, high performance
           crystal lattice filters?
           A. Splitting and tumbling
           B. Tumbling and grinding
           C. Etching and splitting
           D. Etching and grinding

*4AF-5.5   What determine the bandwidth and response shape in a crystal lattice
           filter?
           A. The relative frequencies of the individual crystals
           B. The center frequency chosen for the filter
           C. The amplitude of the RF stage preceding the filter
           D. The amplitude of the signals passing through the filter

SUBELEMENT 4AG — Practical Circuits (10 questions)

4AG-1.1   What is a linear electronic voltage regulator?
- A. A regulator that has a ramp voltage as its output
- B. A regulator in which the pass transistor switches from the "off" state to the "on" state
- C. A regulator in which the control device is switched on or off, with the duty cycle proportional to the line or load conditions
- D. A regulator in which the conduction of a control element is varied in direct proportion to the line voltage or load current

4AG-1.2   What is a switching electronic voltage regulator?
- A. A regulator in which the conduction of a control element is varied in direct proportion to the line voltage or load current
- B. A regulator that provides more than one output voltage
- C. A regulator in which the control device is switched on or off, with the duty cycle proportional to the line or load conditions
- D. A regulator that gives a ramp voltage at its output

4AG-1.3   What device is usually used as a stable reference voltage in a linear voltage regulator?
- A. A Zener diode
- B. A tunnel diode
- C. An SCR
- D. A varactor diode

4AG-1.4   What type of linear regulator is used in applications requiring efficient utilization of the primary power source?
- A. A constant current source
- B. A series regulator
- C. A shunt regulator
- D. A shunt current source

4AG-1.5   What type of linear voltage regulator is used in applications where the load on the unregulated voltage source must be kept constant?
- A. A constant current source
- B. A series regulator
- C. A shunt current source
- D. A shunt regulator

4AG-1.6   To obtain the best temperature stability, what should be the operating voltage of the reference diode in a linear voltage regulator?
- A. Approximately 2.0 volts
- B. Approximately 3.0 volts
- C. Approximately 6.0 volts
- D. Approximately 10.0 volts

4AG-1.7   What is the meaning of the term remote sensing with regard to a linear voltage regulator?
- A. The feedback connection to the error amplifier is made directly to the load
- B. Sensing is accomplished by wireless inductive loops
- C. The load connection is made outside the feedback loop
- D. The error amplifier compares the input voltage to the reference voltage

4AG-1.8 What is a three-terminal regulator?
- A. A regulator that supplies three voltages with variable current
- B. A regulator that supplies three voltages at a constant current
- C. A regulator containing three error amplifiers and sensing transistors
- D. A regulator containing a voltage reference, error amplifier, sensing resistors and transistors, and a pass element

*4AG-1.9 What the important characteristics of a three-terminal regulator?
- A. Maximum and minimum input voltage, minimum output current and voltage
- B. Maximum and minimum input voltage, maximum output current and voltage
- C. Maximum and minimum input voltage, minimum output current and maximum output voltage
- D. Maximum and minimum input voltage, minimum output voltage and maximum output current

4AG-2.1 What is the distinguishing feature of a Class A amplifier?
- A. Output for less than 180 degrees of the signal cycle
- B. Output for the entire 360 degrees of the signal cycle
- C. Output for more than 180 degrees and less than 360 degrees of the signal cycle
- D. Output for exactly 180 degrees of the input signal cycle

4AG-2.2 What class of amplifier is distinguished by the presence of output throughout the entire signal cycle and the input never goes into the cutoff region?
- A. Class A
- B. Class B
- C. Class C
- D. Class D

4AG-2.3 What is the distinguishing characteristic of a Class B amplifier?
- A. Output for the entire input signal cycle
- B. Output for greater than 180 degrees and less than 360 degrees of the input signal cycle
- C. Output for less than 180 degrees of the input signal cycle
- D. Output for 180 degrees of the input signal cycle

4AG-2.4 What class of amplifier is distinguished by the flow of current in the output essentially in 180 degree pulses?
- A. Class A
- B. Class B
- C. Class C
- D. Class D

4AG-2.5 What is a Class AB amplifier?
- A. Output is present for more than 180 degrees but less than 360 degrees of the signal input cycle
- B. Output is present for exactly 180 degrees of the input signal cycle
- C. Output is present for the entire input signal cycle
- D. Output is present for less than 180 degrees of the input signal cycle

4AG-2.6 What is the distinguishing feature of a Class C amplifier?
- A. Output is present for less than 180 degrees of the input signal cycle
- B. Output is present for exactly 180 degrees of the input signal cycle
- C. Output is present for the entire input signal cycle
- D. Output is present for more than 180 degrees but less than 360 degrees of the input signal cycle

4AG-2.7 What class of amplifier is distinguished by the bias being set well beyond cutoff?
- A. Class A
- B. Class B
- C. Class C
- D. Class AB

4AG-2.8 Which class of amplifier provides the highest efficiency?
- A. Class A
- B. Class B
- C. Class C
- D. Class AB

4AG-2.9 Which class of amplifier has the highest linearity and least distortion?
- A. Class A
- B. Class B
- C. Class C
- D. Class AB

4AG-2.10 Which class of amplifier has an operating angle of more than 180 degrees but less than 360 degrees when driven by a sine wave signal?
- A. Class A
- B. Class B
- C. Class C
- D. Class AB

4AG-3.1 What is an L-network?
- A. A network consisting entirely of four inductors
- B. A network consisting of an inductor and a capacitor
- C. A network used to generate a leading phase angle
- D. A network used to generate a lagging phase angle

4AG-3.2 What is a pi-network?
- A. A network consisting entirely of four inductors or four capacitors
- B. A Power Incidence network
- C. An antenna matching network that is isolated from ground
- D. A network consisting of one inductor and two capacitors or two inductors and one capacitor

4AG-3.3 What is a pi-L-network?
- A. A Phase Inverter Load network
- B. A network consisting of two inductors and two capacitors
- C. A network with only three discrete parts
- D. A matching network in which all components are isolated from ground

4AG-3.4 Does the L-, pi-, or pi-L-network provide the greatest harmonic suppression?
A. L-network
B. Pi-network
C. Inverse L-network
D. Pi-L-network

4AG-3.5 What are the three most commonly used networks to accomplish a match between an amplifying device and a transmission line?
A. M-network, pi-network and T-network
B. T-network, M-network and Q-network
C. L-network, pi-network and pi-L-network
D. L-network, M-network and C-network

4AG-3.6 How are networks able to transform one impedance to another?
A. Resistances in the networks substitute for resistances in the load
B. The matching network introduces negative resistance to cancel the resistive part of an impedance
C. The matching network introduces transconductance to cancel the reactive part of an impedance
D. The matching network can cancel the reactive part of an impedance and change the value of the resistive part of an impedance

4AG-3.7 Which type of network offers the greater transformation ratio?
A. L-network
B. Pi-network
C. Constant-K
D. Constant-M

4AG-3.8 Why is the L-network of limited utility in impedance matching?
A. It matches a small impedance range
B. It has limited power handling capabilities
C. It is thermally unstable
D. It is prone to self resonance

4AG-3.9 What is an advantage of using a pi-L-network instead of a pi-network for impedance matching between the final amplifier of a vacuum-tube type transmitter and a multiband antenna?
A. Greater transformation range
B. Higher efficiency
C. Lower losses
D. Greater harmonic suppression

4AG-3.10 Which type of network provides the greatest harmonic suppression?
A. L-network
B. Pi-network
C. Pi-L-network
D. Inverse-Pi network

4AG-4.1 What are the three general groupings of filters?
A. High-pass, low-pass and band-pass
B. Inductive, capacitive and resistive
C. Audio, radio and capacitive
D. Hartley, Colpitts and Pierce

4AG-4.2   What is a constant-K filter?

    A. A filter that uses Boltzmann's constant

    B. A filter whose velocity factor is constant over a wide range of frequencies

    C. A filter whose product of the series- and shunt-element impedances is a constant for all frequencies

    D. A filter whose input impedance varies widely over the design bandwidth

4AG-4.3   What is an advantage of a constant-k filter?

    A. It has high attenuation for signals on frequencies far removed from the passband

    B. It can match impedances over a wide range of frequencies

    C. It uses elliptic functions

    D. The ratio of the cutoff frequency to the trap frequency can be varied

4AG-4.4   What is an m-derived filter?

    A. A filter whose input impedance varies widely over the design bandwidth

    B. A filter whose product of the series- and shunt-element impedances is a constant for all frequencies

    C. A filter whose schematic shape is the letter "M"

    D. A filter that uses a trap to attenuate undesired frequencies too near cutoff for a constant-k filter.

4AG-4.5   What are the distinguishing features of a Butterworth filter?

    A. A filter whose product of the series- and shunt-element impedances is a constant for all frequencies

    B. It only requires capacitors

    C. It has a maximally flat response over its passband

    D. It requires only inductors

4AG-4.6   What are the distinguishing features of a Chebyshev filter?

    A. It has a maximally flat response over its passband

    B. It allows ripple in the passband

    C. It only requires inductors

    D. A filter whose product of the series- and shunt-element impedances is a constant for all frequencies

4AG-4.7   When would it be more desirable to use an m-derived filter over a constant-k filter?

    A. When the response must be maximally flat at one frequency

    B. When you need more attenuation at a certain frequency that is too close to the cut-off frequency for a constant-k filter

    C. When the number of components must be minimized

    D. When high power levels must be filtered

4AG-5.1   What condition must exist for a circuit to oscillate?

    A. It must have a gain of less than 1

    B. It must be neutralized

    C. It must have positive feedback sufficient to overcome losses

    D. It must have negative feedback sufficient to cancel the input

4AG-5.2   What are three major oscillator circuits often used in amateur radio equipment?

    A. Taft, Pierce and negative feedback

    B. Colpitts, Hartley and Taft

    C. Taft, Hartley and Pierce

    D. Colpitts, Hartley and Pierce

4AG-5.3　How is the positive feedback coupled to the input in a Hartley oscillator?
   A. Through a neutralizing capacitor
   B. Through a capacitive divider
   C. Through link coupling
   D. Through a tapped coil

4AG-5.4　How is the positive feedback coupled to the input in a Colpitts oscillator?
   A. Through a tapped coil
   B. Through link coupling
   C. Through a capacitive divider
   D. Through a neutralizing capacitor

4AG-5.5　How is the positive feedback coupled to the input in a Pierce oscillator?
   A. Through a tapped coil
   B. Through link coupling
   C. Through a capacitive divider
   D. Through capacitive coupling

4AG-5.6　Which of the three major oscillator circuits used in amateur radio equipment utilizes a quartz crystal?
   A. Negative feedback
   B. Hartley
   C. Colpitts
   D. Pierce

4AG-5.7　What is the piezoelectric effect?
   A. Mechanical vibration of a crystal by the application of a voltage
   B. Mechanical deformation of a crystal by the application of a magnetic field
   C. The generation of electrical energy by the application of light
   D. Reversed conduction states when a P-N junction is exposed to light

4AG-5.8　What is the major advantage of a Pierce oscillator?
   A. It is easy to neutralize
   B. It doesn't require an LC tank circuit
   C. It can be tuned over a wide range
   D. It has a high output power

4AG-5.9　Which type of oscillator circuit is commonly used in a VFO?
   A. Pierce
   B. Colpitts
   C. Hartley
   D. Negative feedback

4AG-5.10　Why is the Colpitts oscillator circuit commonly used in a VFO?
   A. The frequency is a linear function of the load impedance
   B. It can be used with or without crystal lock-in
   C. It is stable
   D. It has high output power

4AG-6.1 What is meant by the term <u>modulation</u>?
  A. The squelching of a signal until a critical signal-to-noise ratio is reached
  B. Carrier rejection through phase nulling
  C. A linear amplification mode
  D. A mixing process whereby information is imposed upon a carrier

*4AG-6.2 What are the two general categories of methods for generating emission F3E?
  A. The only way to produce an emission F3E signal is with a balanced modulator on the audio amplifier
  B. The only way to produce an emission F3E signal is with a reactance modulator on the oscillator
  C. The only way to produce an emission F3E signal is with a reactance modulator on the final amplifier
  D. The only way to produce an emission F3E signal is with a balanced modulator on the oscillator

4AG-6.3 What is a <u>reactance</u> modulator?
  A. A circuit that acts as a variable resistance or capacitance to produce FM signals
  B. A circuit that acts as a variable resistance or capacitance to produce AM signals
  C. A circuit that acts as a variable inductance or capacitance to produce FM signals
  D. A circuit that acts as a variable inductance or capacitance to produce AM signals

4AG-6.4 What is a <u>balanced</u> modulator?
  A. An FM modulator that produces a balanced deviation
  B. A modulator that produces a double sideband, suppressed carrier signal
  C. A modulator that produces a single sideband, suppressed carrier signal
  D. A modulator that produces a full carrier signal

4AG-6.5 How can an emission J3E signal be generated?
  A. By driving a product detector with a DSB signal
  B. By using a reactance modulator followed by a mixer
  C. By using a loop modulator followed by a mixer
  D. By using a balanced modulator followed by a filter

4AG-6.6 How can an emission A3E signal be generated?
  A. By feeding a phase modulated signal into a low pass filter
  B. By using a balanced modulator followed by a filter
  C. By detuning a Hartley oscillator
  D. By modulating the plate voltage of a class C amplifier

4AG-7.1 How is the efficiency of a power amplifier determined?

  A. $\text{Efficiency} = \dfrac{\text{RF power out}}{\text{dc power in}} \times 100\%$

  B. $\text{Efficiency} = \dfrac{\text{RF power in}}{\text{RF power out}} \times 100\%$

  C. $\text{Efficiency} = \dfrac{\text{RF power in}}{\text{dc power in}} \times 100\%$

  D. $\text{Efficiency} = \dfrac{\text{dc power in}}{\text{RF power in}} \times 100\%$

4AG-7.2   For reasonably efficient operation of a vacuum tube Class C
           amplifier, what should the plate-load resistance be with 1500-volts
           at the plate and 500-milliamperes plate current?
           A. 2000 ohms
           B. 1500 ohms
           C. 4800 ohms
           D. 480 ohms

*4AG-7.3   For reasonably efficient operation of a vacuum Class B amplifier,
           what should the plate-load resistance be with 800-volts at the
           plate and 75-milliamperes plate current?
           A. 679.4 ohms
           B. 60 ohms
           C. 6794 ohms
           D. 10,667 ohms

*4AG-7.4   For reasonably efficient operation of a vacuum tube Class A
           operation what should the plate-load resistance be with 250-volts
           at the plate and 25-milliamperes plate current?
           A. 7692 ohms
           B. 3250 ohms
           C. 325 ohms
           D. 769.2 ohms

4AG-7.5   For reasonably efficient operation of a transistor amplifier, what
           should the load resistance be with 12-volts at the collector and 5
           watts power output?
           A. 100.3 ohms
           B. 14.4 ohms
           C. 10.3 ohms
           D. 144 ohms

4AG-7.6   What is the flywheel effect?
           A. The continued motion of a radio wave through space when
              the transmitter is turned off
           B. The back and forth oscillation of electrons in an LC circuit
           C. The use of a capacitor in a power supply to filter rectified ac
           D. The transmission of a radio signal to a distant station by
              several hops through the ionosphere

4AG-7.7   How can a power amplifier be neutralized?
           A. By increasing the grid drive
           B. By feeding back an in-phase component of the output to the
              input
           C. By feeding back an out-of-phase component of the output to
              the input
           D. By feeding back an out-of-phase component of the input to
              the output

4AG-7.8   What order of Q is required by a tank-circuit sufficient to reduce
           harmonics to an acceptable level?
           A. Approximately 120
           B. Approximately 12
           C. Approximately 1200
           D. Approximately 1.2

4AG-7.9 How can parasitic oscillations be eliminated from a power amplifier?
A. By tuning for maximum SWR
B. By tuning for maximum power output
C. By neutralization
D. By tuning the output

4AG-7.10 What is the procedure for tuning a power amplifier having an output pi-network?
A. Adjust the loading capacitor to maximum capacitance and then dip the plate current with the tuning capacitor
B. Alternately increase the plate current with the tuning capacitor and dip the plate current with the loading capacitor
C. Adjust the tuning capacitor to maximum capacitance and then dip the plate current with the loading capacitor
D. Alternately increase the plate current with the loading capacitor and dip the plate current with the tuning capacitor

4AG-8.1 What is the process of detection?
A. The process of masking out the intelligence on a received carrier to make an S-meter operational
B. The recovery of intelligence from the modulated RF signal
C. The modulation of a carrier
D. The mixing of noise with the received signal

4AG-8.2 What is the principle of detection in a diode detector?
A. Rectification and filtering of RF
B. Breakdown of the Zener voltage
C. Mixing with noise in the transition region of the diode
D. The change of reactance in the diode with respect to frequency

4AG-8.3 What is a product detector?
A. A detector that provides local oscillations for input to the mixer
B. A detector that amplifies and narrows the band-pass frequencies
C. A detector that uses a mixing process with a locally generated carrier
D. A detector used to detect cross-modulation products

4AG-8.4 How are emission F3E signals detected?
A. By a balanced modulator
B. By a frequency discriminator
C. By a product detector
D. By a phase splitter

4AG-8.5 What is a frequency discriminator?
A. A circuit for detecting FM signals
B. A circuit for filtering two closely adjacent signals
C. An automatic bandswitching circuit
D. An FM generator

4AG-8.6 What is the mixing process?
A. The elimination of noise in a wideband receiver by phase comparison
B. The elimination of noise in a wideband receiver by phase differentiation
C. Distortion caused by auroral propagation
D. The combination of two signals to produce sum and difference frequencies

4AG-8.7   What are the principal frequencies which appear at the output of a mixer circuit?
A. Two and four times the original frequency
B. The sum, difference and square root of the input frequencies
C. The original frequencies and the sum and difference frequencies
D. 1.414 and 0.707 times the input frequency

4AG-8.8   What are the advantages of the frequency-conversion process?
A. Automatic squelching and increased selectivity
B. Increased selectivity and optimal tuned-circuit design
C. Automatic soft limiting and automatic squelching
D. Automatic detection in the RF amplifier and increased selectivity

4AG-8.9   What occurs in a receiver when an excessive amount of signal energy reaches the mixer circuit?
A. Spurious mixer products are generated
B. Mixer blanking occurs
C. Automatic limiting occurs
D. A beat frequency is generated

4AG-9.1   How much gain should be used in the rf amplifier stage of a receiver?
A. As much gain as possible short of self oscillation
B. Sufficient gain to allow weak signals to overcome noise generated in the first mixer stage
C. Sufficient gain to keep weak signals below the noise of the first mixer stage
D. It depends on the amplification factor of the first IF stage

4AG-9.2   Why should the rf amplifier stage of a receiver only have sufficient gain to allow weak signals to overcome noise generated in the first mixer stage?
A. To prevent the sum and difference frequencies from being generated
B. To prevent bleed-through of the desired signal
C. To prevent the generation of spurious mixer products
D. To prevent bleed-through of the local oscillator

4AG-9.3   What is the primary purpose of an rf amplifier in a receiver?
A. To provide most of the receiver gain
B. To vary the receiver image rejection by utilizing the AGC
C. To improve the receiver's noise figure
D. To develop the AGC voltage

4AG-9.4   What is an i-f amplifier stage?
A. A fixed-tuned pass-band amplifier
B. A receiver demodulator
C. A receiver filter
D. A buffer oscillator

4AG-9.5   What factors should be considered when selecting an intermediate frequency?
A. Cross-modulation distortion and interference
B. Interference to other services
C. Image rejection and selectivity
D. Noise figure and distortion

4AG-9.6 What is the primary purpose of the first i-f amplifier stage in a receiver?
- A. Gain
- B. Tune out cross-modulation distortion
- C. Dynamic response
- D. Image rejection

4AG-9.7 What is the primary purpose of the final i-f amplifier stage in a receiver?
- A. Sensitivity
- B. Selectivity
- C. Noise figure performance
- D. Squelch gain

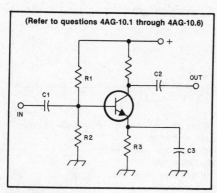

FIGURE 4AG-10

4AG-10.1 What type of circuit is shown in Figure 4AG-10?
- A. Switching voltage regulator
- B. Linear voltage regulator
- C. Common emitter amplifier
- D. Emitter follower amplifier

4AG-10.2 In Figure 4AG-10, what is the purpose of R1 and R2?
- A. Load resistors
- B. Fixed bias
- C. Self bias
- D. Feedback

4AG-10.3 In Figure 4AG-10, what is the purpose of C1?
- A. Decoupling
- B. Output coupling
- C. Self bias
- D. Input coupling

4AG-10.4 In Figure 4AG-10, what is the purpose of C3?
- A. AC feedback
- B. Input coupling
- C. Power supply decoupling
- D. Emitter bypass

4AG-10.5   In Figure 4AG-10, what is the purpose of R3?
      A. Fixed bias
      B. Emitter bypass
      C. Output load resistor
      D. Self bias

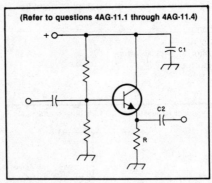

(Refer to questions 4AG-11.1 through 4AG-11.4)

FIGURE 4AG-11

4AG-11.1   What type of circuit is shown in Figure 4AG-11?
      A. High-gain amplifier
      B. Common-collector amplifier
      C. Linear voltage regulator
      D. Grounded-emitter amplifier

4AG-11.2   In Figure 4AG-11, what is the purpose of R?
      A. Emitter load
      B. Fixed bias
      C. Collector load
      D. Voltage regulation

4AG-11.3   In Figure 4AG-11, what is the purpose of C1?
      A. Input coupling
      B. Output coupling
      C. Emitter bypass
      D. Collector bypass

4AG-11.4   In Figure 4AG-11, what is the purpose of C2?
      A. Output coupling
      B. Emitter bypass
      C. Input coupling
      D. Hum filtering

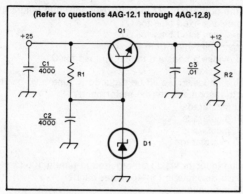

FIGURE 4AG-12

4AG-12.1   What type of circuit is shown in Figure 4AG-12?
    A. Switching voltage regulator
    B. Grounded emitter amplifier
    C. Linear voltage regulator
    D. Emitter follower

4AG-12.2   What is the purpose of D1 in the circuit shown in Figure 4AG-12?
    A. Line voltage stabilization
    B. Voltage reference
    C. Peak clipping
    D. Hum filtering

4AG-12.3   What is the purpose of Q1 in the circuit shown in Figure 4AG-12?
    A. It increases the output ripple
    B. It provides a constant load for the voltage source
    C. It increases the current handling capability
    D. It provides D1 with current

4AG-12.4   What is the purpose of C1 in the circuit shown in Figure 4AG-12?
    A. It resonates at the ripple frequency
    B. It provides fixed bias for Q1
    C. It decouples the output
    D. It filters the supply voltage

4AG-12.5   What is the purpose of C2 in the circuit shown in Figure 4AG-12?
    A. It bypasses hum around D1
    B. It is a brute force filter for the output
    C. To self resonate at the hum frequency
    D. To provide fixed dc bias for Q1

4AG-12.6   What is the purpose of C3 in the circuit shown in Figure 4AG-12?
    A. It prevents self-oscillation
    B. It provides brute force filtering of the output
    C. It provides fixed bias for Q1
    D. It clips the peaks of the ripple

4AG-12.7   What is the purpose of R1 in the circuit shown in Figure 4AG-12?
    A. It provides a constant load to the voltage source
    B. It couples hum to D1
    C. It supplies current to D1
    D. It bypasses hum around D1

4AG-12.8   What is the purpose of R2 in the circuit shown in Figure 4AG-12?
　　A. It provides fixed bias for Q1
　　B. It provides fixed bias for D1
　　C. It decouples hum from D1
　　D. It provides a constant minimum load for Q1

4AG-13.1   What value capacitor would be required to tune a 20-microhenry inductor to resonate in the 80 meter band?
　　A. 150 picofarads
　　B. 200 picofarads
　　C. 100 picofarads
　　D. 100 microfarads

4AG-13.2   What value inductor would be required to tune a 100-picofarad capacitor to resonate in the 40 meter band?
　　A. 200 microhenrys
　　B. 150 microhenrys
　　C. 5 millihenrys
　　D. 5 microhenrys

4AG-13.3   What value capacitor would be required to tune a 2-microhenry inductor to resonate in the 20 meter band?
　　A. 64 picofarads
　　B. 6 picofarads
　　C. 12 picofarads
　　D. 88 microfarads

4AG-13.4   What value inductor would be required to tune a 15-picofarad capacitor to resonate in the 15 meter band?
　　A. 2 microhenrys
　　B. 30 microhenrys
　　C. 4 microhenrys
　　D. 15 microhenrys

4AG-13.5   What value capacitor would be required to tune a 100-microhenry inductor to resonate in the 160 meter band?
　　A. 78 picofarads
　　B. 25 picofarads
　　C. 405 picofarads
　　D. 40.5 microfarads

SUBELEMENT 4AH — Signals and Emissions (6 questions)

4AH-1.1　What is emission A3C?
  A. Facsimile
  B. RTTY
  C. ATV
  D. Slow Scan TV

4AH-1.2　What type of emission is produced when an amplitude modulated transmitter is modulated by a facsimile signal?
  A. A3F
  B. A3C
  C. F3F
  D. F3C

4AH-1.3　What is facsimile?
  A. The transmission of tone-modulated telegraphy
  B. The transmission of a pattern of printed characters designed to form a picture
  C. The transmission of printed pictures by electrical means
  D. The transmission of moving pictures by electrical means

4AH-1.4　What is emission F3C?
  A. Voice transmission
  B. Slow Scan TV
  C. RTTY
  D. Facsimile

4AH-1.5　What type of emission is produced when a frequency modulated transmitter is modulated by a facsimile signal?
  A. F3C
  B. A3C
  C. F3F
  D. A3F

4AH-1.6　What is emission A3F?
  A. RTTY
  B. Television
  C. SSB
  D. Modulated CW

4AH-1.7　What type of emission is produced when an amplitude modulated transmitter is modulated by a television signal?
  A. F3F
  B. A3F
  C. A3C
  D. F3C

4AH-1.8　What is emission F3F?
  A. Modulated CW
  B. Facsimile
  C. RTTY
  D. Television

4AH-1.9　What type of emission is produced when a frequency modulated transmitter is modulated by a television signal?
  A. A3F
  B. A3C
  C. F3F
  D. F3C

4AH-1.10   What type of emission results when a single sideband transmitter
            is used for slow-scan television?
            A. J3A
            B. F3F
            C. A3F
            D. J3F

4AH-2.1   How can an emission F3E signal be produced?
            A. By modulating the supply voltage to a class-B amplifier
            B. By modulating the supply voltage to a class-C amplifier
            C. By using a reactance modulator on an oscillator
            D. By using a balanced modulator on an oscillator

4AH-2.2   How can an emission A3E signal be produced?
            A. By using a reactance modulator on an oscillator
            B. By varying the voltage to the varactor in an oscillator circuit
            C. By using a phase detector, oscillator and filter in a feedback
               loop
            D. By modulating the plate supply voltage to a class C amplifier

4AH-2.3   How can an emission J3E signal be produced?
            A. By producing a double sideband signal with a balanced
               modulator and then removing the unwanted sideband by
               filtering
            B. By producing a double sideband signal with a balanced
               modulator and then removing the unwanted sideband by
               heterodyning
            C. By producing a double sideband signal with a balanced
               modulator and then removing the unwanted sideband by
               mixing
            D. By producing a double sideband signal with a balanced
               modulator and then removing the unwanted sideband by
               neutralization

4AH-3.1   What is meant by the term deviation ratio?
            A. The ratio of the audio modulating frequency to the center
               carrier frequency
            B. The ratio of the maximum carrier frequency deviation to the
               highest audio modulating frequency
            C. The ratio of the carrier center frequency to the audio
               modulating frequency
            D. The ratio of the highest audio modulating frequency to the
               average audio modulating frequency

4AH-3.2   In an emission F3E signal, what is the term for the maximum
            deviation from the carrier frequency divided by the maximum
            audio modulating frequency?
            A. Deviation index
            B. Modulation index
            C. Deviation ratio
            D. Modulation ratio

4AH-3.3   What is the deviation ratio for an emission F3E signal having a
            maximum frequency swing of plus or minus 5 kHz and accepting a
            maximum modulation rate of 3 kHz?
            A. 60
            B. 0.16
            C. 0.6
            D. 1.66

4AH-3.4   What is the deviation ratio for an emission F3E signal having a maximum frequency swing of plus or minus 7.5 kHz and accepting a maximum modulation rate of 3.5 kHz?
   A. 2.14
   B. 0.214
   C. 0.47
   D. 47

4AH-4.1   What is meant by the term modulation index?
   A. The processor index
   B. The ratio between the deviation of a frequency modulated signal and the modulating frequency
   C. The FM signal-to-noise ratio
   D. The ratio of the maximum carrier frequency deviation to the highest audio modulating frequency

4AH-4.2   In an emission F3E signal, what is the term for the ratio between the deviation of a frequency modulated signal and the modulating frequency?
   A. FM compressibility
   B. Quieting index
   C. Percentage of modulation
   D. Modulation index

4AH-4.3   How does the modulation index of a phase-modulated emission vary with the modulated frequency?
   A. The modulation index increases as the RF carrier frequency (the modulated frequency) increases
   B. The modulation index decreases as the RF carrier frequency (the modulated frequency) increases
   C. The modulation index varies with the square root of the RF carrier frequency (the modulated frequency)
   D. The modulation index does not depend on the RF carrier frequency (the modulated frequency)

4AH-4.4   In an emission F3E signal having a maximum frequency deviation of 3000 Hz either side of the carrier frequency, what is the modulation index when the modulating frequency is 1000 Hz?
   A. 3
   B. 0.3
   C. 3000
   D. 1000

4AH-4.5   What is the modulation index of an emission F3E transmitter producing an instantaneous carrier deviation of 6-kHz when modulated with a 2-kHz modulating frequency?
   A. 6000
   B. 3
   C. 2000
   D. 1/3

4AH-5.1   What are electromagnetic waves?
   A. Alternating currents in the core of an electromagnet
   B. A wave consisting of two electric fields at right angles to each other
   C. A wave consisting of an electric field and a magnetic field at right angles to each other
   D. A wave consisting of two magnetic fields at right angles to each other

4AH-5.2   What is a wave front?
          A. A voltage pulse in a conductor
          B. A current pulse in a conductor
          C. A voltage pulse across a resistor
          D. A fixed point in an electromagnetic wave

4AH-5.3   At what speed do electromagnetic waves travel in free space?
          A. Approximately 300 million meters per second
          B. Approximately 468 million meters per second
          C. Approximately 186,300 feet per second
          D. Approximately 300 million miles per second

4AH-5.4   What are the two interrelated fields considered to make up an
          electromagnetic wave?
          A. An electric field and a current field
          B. An electric field and a magnetic field
          C. An electric field and a voltage field
          D. A voltage field and a current field

4AH-5.5   Why do electromagnetic waves not penetrate a good conductor to
          any great extent?
          A. The electromagnetic field induces currents in the insulator
          B. The oxide on the conductor surface acts as a shield
          C. Because of Eddy currents
          D. The resistivity of the conductor dissipates the field

*4AH-6.1 What is meant by referring to electromagnetic waves travel in free
          space?
          A. The electric and magnetic fields eventually become aligned
          B. Propagation in a medium with a high refractive index
          C. The electromagnetic wave encounters the ionosphere and
             returns to its source
          D. Propagation of energy across a vacuum by changing electric
             and magnetic fields

4AH-6.2   What is meant by referring to electromagnetic waves as horizon-
          tally polarized?
          A. The electric field is parallel to the earth
          B. The magnetic field is parallel to the earth
          C. Both the electric and magnetic fields are horizontal
          D. Both the electric and magnetic fields are vertical

4AH-6.3   What is meant by referring to electromagnetic waves as having
          circular polarization?
          A. The electric field is bent into a circular shape
          B. The electric field rotates
          C. The electromagnetic wave continues to circle the earth
          D. The electromagnetic wave has been generated by a quad
             antenna

4AH-6.4   When the electric field is perpendicular to the surface of the
          earth, what is the polarization of the electromagnetic wave?
          A. Circular
          B. Horizontal
          C. Vertical
          D. Elliptical

4AH-6.5 When the magnetic field is parallel to the surface of the earth, what is the polarization of the electromagnetic wave?
A. Circular
B. Horizontal
C. Elliptical
D. Vertical

4AH-6.6 When the magnetic field is perpendicular to the surface of the earth, what is the polarization of the electromagnetic field?
A. Horizontal
B. Circular
C. Elliptical
D. Vertical

4AH-6.7 When the electric field is parallel to the surface of the earth, what is the polarization of the electromagnetic wave?
A. Vertical
B. Horizontal
C. Circular
D. Elliptical

4AH-7.1 What is a sine wave?
A. A constant-voltage, varying-current wave
B. A wave whose amplitude at any given instant can be represented by a point on a wheel rotating at a uniform speed
C. A wave following the laws of the trigonometric tangent function
D. A wave whose polarity changes in a random manner

4AH-7.2 How many times does a sine wave cross the zero axis in one complete cycle?
A. 180 times
B. 4 times
C. 2 times
D. 360 times

4AH-7.3 How many degrees are there in one complete sine wave cycle?
A. 90 degrees
B. 270 degrees
C. 180 degrees
D. 360 degrees

4AH-7.4 What is the period of a wave?
A. The time required to complete one cycle
B. The number of degrees in one cycle
C. The number of zero crossings in one cycle
D. The amplitude of the wave

4AH-7.5 What is a square wave?
A. A wave with only 300 degrees in one cycle
B. A wave which abruptly changes back and forth between two voltage levels and which remains an equal time at each level
C. A wave that makes four zero crossings per cycle
D. A wave in which the positive and negative excursions occupy unequal portions of the cycle time

4AH-7.6   What is a wave called which abruptly changes back and forth
           between two voltage levels and which remains an equal time at
           each level?
           A. A sine wave
           B. A cosine wave
           C. A square wave
           D. A rectangular wave

4AH-7.7   Which sine waves make up a square wave?
           A. 0.707 times the fundamental frequency
           B. The fundamental frequency and all odd and even harmonics
           C. The fundamental frequency and all even harmonics
           D. The fundamental frequency and all odd harmonics

4AH-7.8   What type of wave is made up of sine waves of the fundamental
           frequency and all the odd harmonics?
           A. Square wave
           B. Sine wave
           C. Cosine wave
           D. Tangent wave

4AH-7.9   What is a sawtooth wave?
           A. A wave that alternates between two values and spends an
              equal time at each level
           B. A wave with a straight line rise time faster than the fall time
              (or vice versa)
           C. A wave that produces a phase angle tangent to the unit
              circle
           D. A wave whose amplitude at any given instant can be
              represented by a point on a wheel rotating at a uniform
              speed

4AH-7.10   What type of wave is characterized by a rise time significantly
           faster than the fall time (or vice versa)?
           A. A cosine wave
           B. A square wave
           C. A sawtooth wave
           D. A sine wave

4AH-7.11   Which sine waves make up a sawtooth wave?
           A. The fundamental frequency and all prime harmonics
           B. The fundamental frequency and all even harmonics
           C. The fundamental frequency and all odd harmonics
           D. The fundamental frequency and all harmonics

4AH-7.12   What type of wave is made up of sine waves at the fundamental
           frequency and all the harmonics?
           A. A sawtooth wave
           B. A square wave
           C. A sine wave
           D. A cosine wave

4AH-8.1 What is the meaning of the term root mean square value of an ac voltage?
A. The value of an ac voltage found by squaring the average value of the peak ac voltage
B. The value of a dc voltage that would cause the same heating effect in a given resistor as a peak ac voltage
C. The value of an ac voltage that would cause the same heating effect in a given resistor as a dc voltage of the same value
D. The value of an ac voltage found by taking the square root of the average ac value

4AH-8.2 What is the term used in reference to a dc voltage that would cause the same heating in a resistor as a certain value of ac voltage?
A. Cosine voltage
B. Power factor
C. Root mean square
D. Average voltage

4AH-8.3 What would be the most accurate way of determining the rms voltage of a complex waveform?
A. By using a grid dip meter
B. By measuring the voltage with a D'Arsonval meter
C. By using an absorption wavemeter
D. By measuring the heating effect in a known resistor

4AH-8.4 What is the rms voltage at a common household electrical power outlet?
A. 117-V ac
B. 331-V ac
C. 82.7-V ac
D. 165.5-V ac

4AH-8.5 What is the peak voltage at a common household electrical outlet?
A. 234 volts
B. 165.5 volts
C. 117 volts
D. 331 volts

4AH-8.6 What is the peak-to-peak voltage at a common household electrical outlet?
A. 234 volts
B. 117 volts
C. 331 volts
D. 165.5 volts

4AH-8.7 What is the rms voltage of a 165-volt peak pure sine wave?
A. 233-V ac
B. 330-V ac
C. 58.3-V ac
D. 117-V ac

4AH-8.8 What is the rms value of a 331-volt peak-to-peak pure sine wave?
A. 117-V ac
B. 165-V ac
C. 234-V ac
D. 300-V ac

4AH-9.1 For many types of voices, what is the ratio of PEP to average power during a modulation peak in an emission J3E signal?
A. Approximately 1.0 to 1
B. Approximately 25 to 1
C. Approximately 2.5 to 1
D. Approximately 100 to 1

4AH-9.2 In an emission J3E signal, what determines the PEP-to-average power ratio?
A. The frequency of the modulating signal
B. The degree of carrier suppression
C. The speech characteristics
D. The amplifier power

4AH-9.3 What is the approximate dc input power to a Class B rf power amplifier stage in an emission F3E transmitter when the PEP output power is 1500 watts?
A. Approximately 900 watts
B. Approximately 1765 watts
C. Approximately 2500 watts
D. Approximately 3000 watts

4AH-9.4 What is the approximate dc input power to a Class C rf power amplifier stage in an emission F1B transmitter when the PEP output power is 1000 watts?
A. Approximately 850 watts
B. Approximately 1250 watts
C. Approximately 1667 watts
D. Approximately 2000 watts

4AH-9.5 What is the approximate dc input power to a Class AB rf power amplifier stage in an emission N0N transmitter when the PEP output power is 500 watts?
A. Approximately 250 watts
B. Approximately 600 watts
C. Approximately 800 watts
D. Approximately 1000 watts

4AH-10.1 Where is the noise generated which primarily determines the signal-to-noise ratio in a 160 meter band receiver?
A. In the detector
B. Man-made noise
C. In the receiver front end
D. In the atmosphere

4AH-10.2 Where is the noise generated which primarily determines the signal-to-noise ratio in a 2 meter band receiver?
A. In the receiver front end
B. Man-made noise
C. In the atmosphere
D. In the ionosphere

4AH-10.3 Where is the noise generated which primarily determines the signal-to-noise ratio in a 1.25 meter band receiver?
A. In the audio amplifier
B. In the receiver front end
C. In the ionosphere
D. Man-made noise

4AH-10.4 Where is the noise generated which primarily determines the signal-to-noise ratio in a 0.70 meter band receiver?
- A. In the atmosphere
- B. In the ionosphere
- C. In the receiver front end
- D. Man-made noise

4AI-1.1   What is meant by the term <u>antenna gain</u>?
- A. The numerical ratio relating the radiated signal strength of an antenna to that of another antenna
- B. The ratio of the signal in the forward direction to the signal in the back direction
- C. The ratio of the amount of power produced by the antenna compared to the output power of the transmitter
- D. The final amplifier gain minus the transmission line losses (including any phasing lines present)

4AI-1.2   What is the term for a numerical ratio which relates the performance of one antenna to that of another real or theoretical antenna?
- A. Effective radiated power
- B. Antenna gain
- C. Conversion gain
- D. Peak effective power

4AI-1.3   What is meant by the term <u>antenna bandwidth</u>?
- A. Antenna length divided by the number of elements
- B. The frequency range over which an antenna can be expected to perform well
- C. The angle between the half-power radiation points
- D. The angle formed between two imaginary lines drawn through the ends of the elements

4AI-1.4   How can the approximate beamwidth of a rotatable beam antenna be determined?
- A. Note the two points where the signal strength of the antenna is down 3 dB from the maximum signal point and compute the angular difference
- B. Measure the ratio of the signal strengths of the radiated power lobes from the front and rear of the antenna
- C. Draw two imaginary lines through the ends of the elements and measure the angle between the lines
- D. Measure the ratio of the signal strengths of the radiated power lobes from the front and side of the antenna

4AI-2.1   What is a <u>trap antenna</u>?
- A. An antenna for rejecting interfering signals
- B. A highly sensitive antenna with maximum gain in all directions
- C. An antenna capable of being used on more than one band because of the presence of parallel LC networks
- D. An antenna with a large capture area

4AI-2.2   What is an advantage of using a trap antenna?
- A. It has high directivity in the high-frequency amateur bands
- B. It has high gain
- C. It minimizes harmonic radiation
- D. It may be used for multiband operation

4AI-2.3   What is a disadvantage of using a trap antenna?
- A. It will radiate harmonics
- B. It can only be used for single band operation
- C. It is too sharply directional at the lower amateur frequencies
- D. It must be neutralized

4AI-2.4   What is the principle of a trap antenna?
           A. Beamwidth may be controlled by non-linear impedances
           B. The traps form a high impedance to isolate parts of the
              antenna
           C. The effective radiated power can be increased if the space
              around the antenna "sees" a high impedance
           D. The traps increase the antenna gain

4AI-3.1   What is a parasitic element of an antenna?
           A. An element polarized 90 degrees opposite the driven
              element
           B. An element dependent on the antenna structure for support
           C. An element that receives its excitation from mutual coupling
              rather than from a transmission line
           D. A transmission line that radiates radio-frequency energy

4AI-3.2   How does a parasitic element generate an electromagnetic field?
           A. By the RF current received from a connected transmission
              line
           B. By interacting with the earth's magnetic field
           C. By altering the phase of the current on the driven element
           D. By currents induced into the element from a surrounding
              electric field

4AI-3.3   How does the length of the reflector element of a parasitic
           element beam antenna compare with that of the driven element?
           A. It is about 5% longer
           B. It is about 5% shorter
           C. It is twice as long
           D. It is one-half as long

4AI-3.4   How does the length of the director element of a parasitic element
           beam antenna compare with that of the driven element?
           A. It is about 5% longer
           B. It is about 5% shorter
           C. It is one-half as long
           D. It is twice as long

4AI-4.1   What is meant by the term radiation resistance for an antenna?
           A. Losses in the antenna elements and feed line
           B. The specific impedance of the antenna
           C. An equivalent resistance that would dissipate the same
              amount of power as that radiated from an antenna
           D. The resistance in the trap coils to received signals

4AI-4.2   What is the term used for an equivalent resistance which would
           dissipate the same amount of energy as that radiated from an
           antenna?
           A. Space resistance
           B. Loss resistance
           C. Transmission line loss
           D. Radiation resistance

4AI-4.3 Why is the value of the radiation resistance of an antenna important?
  A. Knowing the radiation resistance makes it possible to match impedances for maximum power transfer
  B. Knowing the radiation resistance makes it possible to measure the near-field radiation density from a transmitting antenna
  C. The value of the radiation resistance represents the front-to-side ratio of the antenna
  D. The value of the radiation resistance represents the front-to-back ratio of the antenna

4AI-4.4 What are the factors that determine the radiation resistance of an antenna?
  A. Transmission line length and height of an antenna
  B. The location of the antenna with respect to nearby objects and the length/diameter ratio of the conductors
  C. It is a constant for all antennas since it is a physical constant
  D. Sunspot activity and the time of day

4AI-5.1 What is a driven element of an antenna?
  A. Always the rearmost element
  B. Always the forwardmost element
  C. The element fed by the transmission line
  D. The element connected to the rotator

4AI-5.2 What is the usual electrical length of a driven element in a HF beam antenna?
  A. 1/4 wavelength
  B. 1/2 wavelength
  C. 3/4 wavelength
  D. 1 wavelength

4AI-5.3 What is the term for an antenna element which is supplied power from a transmitter through a transmission line?
  A. Driven element
  B. Director element
  C. Reflector element
  D. Parasitic element

4AI-6.1 What is meant by the term antenna efficiency?

  A. $\text{Efficiency} = \dfrac{\text{radiation resistance}}{\text{transmission resistance}} \times 100\%$

  B. $\text{Efficiency} = \dfrac{\text{radiation resistance}}{\text{total resistance}} \times 100\%$

  C. $\text{Efficiency} = \dfrac{\text{total resistance}}{\text{radiation resistance}} \times 100\%$

  D. $\text{Efficiency} = \dfrac{\text{effective radiated power}}{\text{transmitter output}} \times 100\%$

4AI-6.2 What is the term for the ratio of the radiation resistance of an antenna to the total resistance of the system?
  A. Effective radiated power
  B. Radiation conversion loss
  C. Antenna efficiency
  D. Beamwidth

4AI-6.3 What is included in the total resistance of an antenna system?
  A. Radiation resistance plus space impedance
  B. Radiation resistance plus transmission resistance
  C. Transmission line resistance plus radiation resistance
  D. Radiation resistance plus ohmic resistance

4AI-6.4 How can the antenna efficiency of a HF grounded vertical antenna be made comparable to that of a half-wave antenna?
  A. By installing a good ground radial system
  B. By isolating the coax shield from ground
  C. By shortening the vertical
  D. By lengthening the vertical

4AI-6.5 Why does a half-wave antenna operate at very high efficiency?
  A. Because it is non-resonant
  B. Because the conductor resistance is low compared to the radiation resistance
  C. Because earth-induced currents add to its radiated power
  D. Because it has less corona from the element ends than other types of antennas

4AI-7.1 What is a folded dipole antenna?
  A. A dipole that is one-quarter wavelength long
  B. A ground plane antenna
  C. A dipole whose ends are connected by another one-half wavelength piece of wire
  D. A fictional antenna used in theoretical discussions to replace the radiation resistance

4AI-7.2 How does the bandwidth of a folded dipole antenna compare with that of a simple dipole antenna?
  A. It is 0.707 times the simple dipole bandwidth
  B. It is essentially the same
  C. It is less than 50% that of a simple dipole
  D. It is greater

4AI-7.3 What is the input terminal impedance at the center of a folded dipole antenna?
  A. 300 ohms
  B. 72 ohms
  C. 50 ohms
  D. 450 ohms

4AI-8.1 What is the meaning of the term velocity factor of a transmission line?
  A. The ratio of the characteristic impedance of the line to the terminating impedance
  B. The index of shielding for coaxial cable
  C. The velocity of the wave on the transmission line multiplied by the velocity of light in a vacuum
  D. The velocity of the wave on the transmission line divided by the velocity of light in a vacuum

4AI-8.2   What is the term for the ratio of actual velocity at which a signal
           travels through a line to the speed of light in a vacuum?
           A. Velocity factor
           B. Characteristic impedance
           C. Surge impedance
           D. Standing wave ratio

4AI-8.3   What is the velocity factor for a typical coaxial cable?
           A. 2.70
           B. 0.66
           C. 0.30
           D. 0.10

4AI-8.4   What determines the velocity factor in a transmission line?
           A. The termination impedance
           B. The line length
           C. Dielectrics in the line
           D. The center conductor resistivity

4AI-8.5   Why is the physical length of a coaxial cable transmission line
           shorter than its electrical length?
           A. Skin effect is less pronounced in the coaxial cable
           B. RF energy moves slower along the coaxial cable
           C. The surge impedance is higher in the parallel feed line
           D. The characteristic impedance is higher in the parallel feed line

4AI-9.1   What would be the physical length of a typical coaxial transmis-
           sion line which is electrically one-quarter wavelength long at
           14.1 MHz?
           A. 20 meters
           B. 3.55 meters
           C. 2.51 meters
           D. 0.25 meters

4AI-9.2   What would be the physical length of a typical coaxial transmis-
           sion line which is electrically one-quarter wavelength long at
           7.2 MHz?
           A. 10.5 meters
           B. 6.88 meters
           C. 24 meters
           D. 50 meters

4AI-9.3   What is the physical length of a parallel antenna feedline which is
           electrically one-half wavelength long at 14.10 MHz? (assume a
           velocity factor of 0.82.)
           A. 15 meters
           B. 24.3 meters
           C. 8.7 meters
           D. 70.8 meters

4AI-9.4   What is the physical length of a twin lead transmission feedline at
           3.65 MHz? (assume a velocity factor of 0.80.)
           A. Electrical length times 0.8
           B. Electrical length divided by 0.8
           C. 80 meters
           D. 160 meters

4AI-10.1   In a half-wave antenna, where are the current nodes?
- A. At the ends
- B. At the feed points
- C. Three-quarters of the way from the feed point toward the end
- D. One-half of the way from the feed point toward the end

4AI-10.2   In a half-wave antenna, where are the voltage nodes?
- A. At the ends
- B. At the feed point
- C. Three-quarters of the way from the feed point toward the end
- D. One-half of the way from the feed point toward the end

4AI-10.3   At the ends of a half-wave antenna, what values of current and voltage exist compared to the remainder of the antenna?
- A. Equal voltage and current
- B. Minimum voltage and maximum current
- C. Maximum voltage and minimum current
- D. Minimum voltage and minimum current

4AI-10.4   At the center of a half-wave antenna, what values of voltage and current exist compared to the remainder of the antenna?
- A. Equal voltage and current
- B. Maximum voltage and minimum current
- C. Minimum voltage and minimum current
- D. Minimum voltage and maximum current

4AI-11.1   Why is the inductance required for a base loaded HF mobile antenna less than that for an inductance placed further up the whip?
- A. The capacitance to ground is less farther away from the base
- B. The capacitance to ground is greater farther away from the base
- C. The current is greater at the top
- D. The voltage is less at the top

4AI-11.2   What happens to the base feed point of a fixed length HF mobile antenna as the frequency of operation is lowered?
- A. The resistance decreases and the capacitive reactance decreases
- B. The resistance decreases and the capacitive reactance increases
- C. The resistance increases and the capacitive reactance decreases
- D. The resistance increases and the capacitive reactance increases

4AI-11.3   Why should an HF mobile antenna loading coil have a high ratio of reactance to resistance?
- A. To swamp out harmonics
- B. To maximize losses
- C. To minimize losses
- D. To minimize the Q

4AI-11.4   Why is a loading coil often used with an HF mobile antenna?
- A. To improve reception
- B. To lower the losses
- C. To lower the Q
- D. To tune out the capacitive reactance

4AI-12.1   For a shortened vertical antenna, where should a loading coil be placed to minimize losses and produce the most effective performance?
A. Near the center of the vertical radiator
B. As low as possible on the vertical radiator
C. As close to the transmitter as possible
D. At a voltage node

4AI-12.2   What happens to the bandwidth of an antenna as it is shortened through the use of loading coils?
A. It is increased
B. It is decreased
C. No change occurs
D. It becomes flat

4AI-12.3   Why are self-resonant antennas popular in amateur stations?
A. They are very broad banded
B. They have high gain in all azimuthal directions
C. They are the most efficient radiators
D. They require no calculations

4AI-12.4   What is an advantage of using top loading in a shortened HF vertical antenna?
A. Lower Q
B. Greater structural strength
C. Higher losses
D. Improved radiation efficiency

Element 4A Answer Key

SUBELEMENT 4AA

4AA-1.1	A
4AA-1.2	B
4AA-1.3	D
4AA-1.4	C
4AA-2.1	A
4AA-2.2	D
4AA-2.3	B
4AA-2.4	A
4AA-3.1	D
4AA-3.2	A
4AA-3.3	C
4AA-3.4	D
4AA-3.5	C
4AA-3.6	A
4AA-3.7	D
4AA-3.8	A
4AA-3.9	B
4AA-3.10	A
4AA-4.1	D
4AA-4.2	A
4AA-4.3	B
4AA-4.4	C
4AA-5.1	D
4AA-5.2	A
4AA-5.3	C
4AA-5.4	C
4AA-5.5	D
4AA-6.1	A
4AA-6.2	B
4AA-6.3	B
4AA-7.1	C
4AA-7.2	D
4AA-8.1	A
4AA-8.2	B
4AA-9.1	C
4AA-9.2	C
4AA-9.3	D
4AA-9.4	A
4AA-10.1	B
4AA-10.2	C
4AA-11.1	B
4AA-11.2	A
4AA-12.1	B
4AA-12.2	C
4AA-12.3	D
4AA-13.1	D
4AA-13.2	B

4AA-14.1	C
4AA-14.2	D
4AA-15.1	A
4AA-15.2	B
4AA-15.3	A
4AA-16.1	C
4AA-16.2	D
4AA-17.1	A
4AA-17.2	B
4AA-17.3	C
4AA-18.1	B
4AA-18.2	D
4AA-18.3	B
4AA-19.1	C
4AA-19.2	A
4AA-19.3	A
4AA-19.4	B
4AA-20.1	C
4AA-20.2	D

SUBELEMENT 4AB

4AB-1.1	D
4AB-1.2	A
4AB-1.3	B
4AB-1.4	B
4AB-1.5	C
4AB-2.1	D
4AB-2.2	B
4AB-2.3	C
4AB-2.4	C
4AB-2.5	D

SUBELEMENT 4AC

4AC-1.1	C
4AC-1.2	D
4AC-1.3	A
4AC-1.4	B
4AC-1.5	A
4AC-2.1	B
4AC-2.2	C
4AC-2.3	D
4AC-2.4	B
4AC-2.5	A
4AC-3.1	D
4AC-3.2	C
4AC-3.3	B
4AC-3.4	D
4AC-3.5	A

4AC-4.1	D	4AE-2.3	D
4AC-4.2	A	4AE-2.4	B
4AC-4.3	B	4AE-2.5	A
4AC-4.4	C	4AE-2.6	B
4AC-4.5	A	4AE-2.7	B
		4AE-3.1	A
SUBELEMENT 4AD		4AE-3.2	C
4AD-1.1	B	4AE-3.3	A
4AD-1.2	A	4AE-3.4	A
4AD-1.3	B	4AE-3.5	C
4AD-1.4	A	4AE-4.1	B
4AD-1.5	D	4AE-4.2	D
4AD-1.6	C	4AE-4.3	C
4AD-1.7	A	4AE-4.4	B
4AD-1.8	D	4AE-4.5	B
4AD-1.9	D	4AE-4.6	A
4AD-1.10	A	4AE-4.7	D
4AD-1.11	C	4AE-5.1	C
4AD-2.1	C	4AE-5.2	B
4AD-2.2	D	4AE-5.3	C
4AD-2.3	B	4AE-5.4	A
4AD-2.4	D	4AE-5.5	B
4AD-2.5	B	4AE-5.6	D
4AD-2.6	A	4AE-5.7	C
4AD-2.7	B	4AE-5.8	A
4AD-3.1	A	4AE-5.9	B
4AD-3.2	D	4AE-5.10	C
4AD-3.3	B	4AE-5.11	A
4AD-3.4	D	4AE-5.12	B
4AD-3.5	C	4AE-5.13	C
4AD-4.1	D	4AE-5.14	D
4AD-4.2	B	4AE-5.15	A
4AD-4.3	B	4AE-5.16	B
4AD-4.4	D	4AE-5.17	C
4AD-4.5	B	4AE-5.18	D
4AD-5.1	C	4AE-5.19	A
4AD-5.2	A	4AE-5.20	B
4AD-5.3	C	4AE-5.21	A
4AD-5.4	C	4AE-5.22	D
4AD-5.5	A	4AE-5.23	C
4AD-6.1	D	4AE-5.24	D
4AD-6.2	B	4AE-5.25	A
4AD-6.3	A	4AE-5.26	D
4AD-6.4	C	4AE-5.27	B
4AD-7.1	C	4AE-5.28	A
4AD-7.2	C	4AE-5.29	C
4AD-7.3	A	4AE-5.30	D
		4AE-5.31	A
SUBELEMENT 4AE		4AE-5.32	B
4AE-1.1	A	4AE-5.33	C
4AE-1.2	D	4AE-5.34	D
4AE-1.3	A	4AE-5.35	D
4AE-1.4	B	4AE-5.36	A
4AE-2.1	C	4AE-5.37	B
4AE-2.2	B	4AE-5.38	B

4AE-5.39	D		4AF-1.15	B
4AE-5.40	A		4AF-1.16	D
4AE-6.1	A		4AF-1.17	C
4AE-6.2	B		4AF-1.18	D
4AE-6.3	C		4AF-1.19	C
4AE-6.4	B		4AF-1.20	C
4AE-6.5	D		4AF-2.1	C
4AE-6.6	B		4AF-2.2	B
4AE-6.7	A		4AF-2.3	B
4AE-6.8	D		4AF-2.4	C
4AE-6.9	D		4AF-2.5	C
4AE-6.10	C		4AF-2.6	A
4AE-7.1	A		4AF-2.7	B
4AE-7.2	A		4AF-2.8	B
4AE-7.3	C		4AF-2.9	B
4AE-7.4	D		4AF-2.10	B
4AE-7.5	C		4AF-2.11	A
4AE-7.6	B		4AF-2.12	A
4AE-7.7	D		4AF-2.13	C
4AE-8.1	B		4AF-2.14	C
4AE-8.2	C		4AF-2.15	A
4AE-8.3	D		4AF-2.16	A
4AE-8.4	A		4AF-2.17	B
4AE-8.5	D		4AF-3.1	D
4AE-8.6	B		4AF-3.2	A
4AE-8.7	C		4AF-3.3	A
4AE-8.8	D		4AF-3.4	A
4AE-8.9	A		4AF-3.5	D
4AE-8.10	D		4AF-3.6	A
4AE-9.1	B		4AF-3.7	A
4AE-9.2	C		4AF-3.8	B
4AE-9.3	C		4AF-4.1	B
4AE-9.4	D		4AF-4.2	C
4AE-9.5	C		4AF-4.3	B
4AE-9.6	A		4AF-4.4	A
4AE-9.7	B		4AF-4.5	D
4AE-9.8	B		4AF-4.6	C
4AE-9.9	C		4AF-4.7	B
4AE-9.10	C		4AF-4.8	A
			4AF-4.9	D
SUBELEMENT 4AF			4AF-4.10	D
4AF-1.1	D		4AF-5.1	B
4AF-1.2	A		4AF-5.2	C
4AF-1.3	D		4AF-5.3	D
4AF-1.4	C		4AF-5.4	D
4AF-1.5	B		4AF-5.5	A
4AF-1.6	A			
4AF-1.7	C		**SUBELEMENT 4AG**	
4AF-1.8	C		4AG-1.1	D
4AF-1.9	C		4AG-1.2	C
4AF-1.10	D		4AG-1.3	A
4AF-1.11	A		4AG-1.4	B
4AF-1.12	B		4AG-1.5	D
4AF-1.13	D		4AG-1.6	C
4AF-1.14	D		4AG-1.7	A

4AG-1.8	D		4AG-8.1	B
4AG-1.9	B		4AG-8.2	A
4AG-2.1	B		4AG-8.3	C
4AG-2.2	A		4AG-8.4	B
4AG-2.3	D		4AG-8.5	A
4AG-2.4	B		4AG-8.6	D
4AG-2.5	A		4AG-8.7	C
4AG-2.6	A		4AG-8.8	B
4AG-2.7	C		4AG-8.9	A
4AG-2.8	C		4AG-9.1	B
4AG-2.9	A		4AG-9.2	C
4AG-2.10	D		4AG-9.3	C
4AG-3.1	B		4AG-9.4	A
4AG-3.2	D		4AG-9.5	C
4AG-3.3	B		4AG-9.6	D
4AG-3.4	D		4AG-9.7	B
4AG-3.5	C		4AG-10.1	C
4AG-3.6	D		4AG-10.2	B
4AG-3.7	B		4AG-10.3	D
4AG-3.8	A		4AG-10.4	D
4AG-3.9	D		4AG-10.5	D
4AG-3.10	C		4AG-11.1	B
4AG-4.1	A		4AG-11.2	A
4AG-4.2	C		4AG-11.3	D
4AG-4.3	A		4AG-11.4	A
4AG-4.4	D		4AG-12.1	C
4AG-4.5	C		4AG-12.2	B
4AG-4.6	B		4AG-12.3	C
4AG-4.7	B		4AG-12.4	D
4AG-5.1	C		4AG-12.5	A
4AG-5.2	D		4AG-12.6	A
4AG-5.3	D		4AG-12.7	C
4AG-5.4	C		4AG-12.8	D
4AG-5.5	D		4AG-13.1	C
4AG-5.6	D		4AG-13.2	D
4AG-5.7	A		4AG-13.3	A
4AG-5.8	B		4AG-13.4	C
4AG-5.9	B		4AG-13.5	A
4AG-5.10	C			
4AG-6.1	D			
4AG-6.2	B		**SUBELEMENT 4AH**	
4AG-6.3	C		4AH-1.1	A
4AG-6.4	B		4AH-1.2	B
4AG-6.5	D		4AH-1.3	C
4AG-6.6	D		4AH-1.4	D
4AG-7.1	A		4AH-1.5	A
4AG-7.2	B		4AH-1.6	B
4AG-7.3	C		4AH-1.7	B
4AG-7.4	A		4AH-1.8	D
4AG-7.5	B		4AH-1.9	C
4AG-7.6	B		4AH-1.10	D
4AG-7.7	C		4AH-2.1	C
4AG-7.8	B		4AH-2.2	D
4AG-7.9	C		4AH-2.3	A
4AG-7.10	D		4AH-3.1	B
			4AH-3.2	C

4AH-3.3	D	**SUBELEMENT 4AI**	
4AH-3.4	A	4AI-1.1	A
4AH-4.1	B	4AI-1.2	B
4AH-4.2	D	4AI-1.3	B
4AH-4.3	D	4AI-1.4	A
4AH-4.4	A	4AI-2.1	C
4AH-4.5	B	4AI-2.2	D
4AH-5.1	C	4AI-2.3	A
4AH-5.2	D	4AI-2.4	B
4AH-5.3	A	4AI-3.1	C
4AH-5.4	B	4AI-3.2	D
4AH-5.5	C	4AI-3.3	A
4AH-6.1	D	4AI-3.4	B
4AH-6.2	A	4AI-4.1	C
4AH-6.3	B	4AI-4.2	D
4AH-6.4	C	4AI-4.3	A
4AH-6.5	D	4AI-4.4	B
4AH-6.6	A	4AI-5.1	C
4AH-6.7	B	4AI-5.2	B
4AH-7.1	B	4AI-5.3	A
4AH-7.2	C	4AI-6.1	B
4AH-7.3	D	4AI-6.2	C
4AH-7.4	A	4AI-6.3	D
4AH-7.5	B	4AI-6.4	A
4AH-7.6	C	4AI-6.5	B
4AH-7.7	D	4AI-7.1	C
4AH-7.8	A	4AI-7.2	D
4AH-7.9	B	4AI-7.3	A
4AH-7.10	C	4AI-8.1	D
4AH-7.11	D	4AI-8.2	A
4AH-7.12	A	4AI-8.3	B
4AH-8.1	C	4AI-8.4	C
4AH-8.2	C	4AI-8.5	B
4AH-8.3	D	4AI-9.1	B
4AH-8.4	A	4AI-9.2	B
4AH-8.5	B	4AI-9.3	C
4AH-8.6	C	4AI-9.4	A
4AH-8.7	D	4AI-10.1	A
4AH-8.8	A	4AI-10.2	B
4AH-9.1	C	4AI-10.3	C
4AH-9.2	C	4AI-10.4	D
4AH-9.3	C	4AI-11.1	A
4AH-9.4	B	4AI-11.2	B
4AH-9.5	D	4AI-11.3	C
4AH-10.1	D	4AI-11.4	D
4AH-10.2	A	4AI-12.1	A
4AH-10.3	B	4AI-12.2	B
4AH-10.4	C	4AI-12.3	C
		4AI-12.4	D

Appendix A

U S Customary—Metric Conversion Factors

International System of Units (SI) — Metric Units

Prefix	Symbol	Multiplication Factor	
exa	E	10^{18} =	1,000,000,000,000,000,000
peta	P	10^{15} =	1,000,000,000,000,000
tera	T	10^{12} =	1,000,000,000,000
giga	G	10^9 =	1,000,000,000
mega	M	10^6 =	1,000,000
kilo	k	10^3 =	1,000
hecto	h	10^2 =	100
deca	da	10^1 =	10
(unit)		10^0 =	1
deci	d	10^{-1} =	0.1
centi	c	10^{-2} =	0.01
milli	m	10^{-3} =	0.001
micro	μ	10^{-6} =	0.000001
nano	n	10^{-9} =	0.000000001
pico	p	10^{-12} =	0.000000000001
femto	f	10^{-15} =	0.000000000000001
atto	a	10^{-18} =	0.000000000000000001

Linear
1 meter (m) = 100 centimeters (cm) = 1000 millimeters (mm)

Area
$1 \text{ m}^2 = 1 \times 10^4 \text{ cm}^2 = 1 \times 10^6 \text{ mm}^2$

Volume
$1 \text{ m}^3 = 1 \times 10^6 \text{ cm}^3 = 1 \times 10^9 \text{ mm}^3$
1 liter (l) = 1000 cm^3 = 1×10^6 mm^3

Mass
1 kilogram (kg) = 1000 grams (g)
  (Approximately the mass of 1 liter of water)
1 metric ton (or tonne) = 1000 kg

U S Customary Units

Linear Units
12 inches (in) = 1 foot (ft)
36 inches = 3 feet = 1 yard (yd)
1 rod = 5½ yards = 16½ feet
1 statute mile = 1760 yards = 5280 feet
1 nautical mile = 6076.11549 feet

Area
1 ft^2 = 144 in^2
1 yd^2 = 9 ft^2 = 1296 in^2
1 rod^2 = 30¼ yd^2
1 acre = 4840 yd^2 = 43,560 ft^2
1 acre = 160 rod^2
1 mile^2 = 640 acres

Volume
1 ft^3 = 1728 in^3
1 yd^3 = 27 ft^3

Liquid Volume Measure
1 fluid ounce (fl oz) = 8 fluidrams = 1.804 in^3
1 pint (pt) = 16 fl oz
1 quart (qt) = 2 pt = 32 fl oz = 57¾ in^3
1 gallon (gal) = 4 qt = 231 in^3
1 barrel = 31½ gal

Dry Volume Measure
1 quart (qt) = 2 pints (pt) = 67.2 in^3
1 peck = 8 qt

1 bushel = 4 pecks = 2150.42 in^3

Avoirdupois Weight
1 dram (dr) = 27.343 grains (gr) or (gr a)
1 ounce (oz) = 437.5 gr
1 pound (lb) = 16 oz = 7000 gr
1 short ton = 2000 lb, 1 long ton = 2240 lb

Troy Weight
1 grain troy (gr t) = 1 grain avoirdupois
1 pennyweight (dwt) or (pwt) = 24 gr t
1 ounce troy (oz t) = 480 grains
1 lb t = 12 oz t = 5760 grains

Apothecaries' Weight
1 grain apothecaries' (gr ap) = 1 gr t = 1 gr a
1 dram ap (dr ap) = 60 gr
1 oz ap = 1 oz t = 8 dr ap = 480 gr
1 lb ap = 1 lb t = 12 oz ap = 5760 gr

Temperature
°F = 9/5 °C + 32
°C = 5/9 (°F − 32)
K = °C + 273
°C = K − 273

Multiply ⟶

Metric Unit = Conversion Factor × U.S. Customary Unit

⟵ Divide

Metric Unit ÷ Conversion Factor = U.S. Customary Unit

Conversion — Metric Unit = Factor × U.S. Unit

Metric Unit	Factor	U.S. Unit		Metric Unit	Factor	U.S. Unit
(Length)				**(Volume)**		
mm	25.4	inch		mm³	16387.064	in³
cm	2.54	inch		cm³	16.387	in³
cm	30.48	foot		m³	0.028316	ft³
m	0.3048	foot		m³	0.764555	yd³
m	0.9144	yard		ml	16.387	in³
km	1.609	mile		ml	29.57	fl oz
km	1.852	nautical mile		ml	473	pint
(Area)				ml	946.333	quart
mm²	645.16	inch²		l	28.32	ft³
cm²	6.4516	in²		l	0.9463	quart
cm²	929.03	ft²		l	3.785	gallon
m²	0.0929	ft²		l	1.101	dry quart
cm²	8361.3	yd²		l	8.809	peck
m²	0.83613	yd²		l	35.238	bushel
m²	4047	acre				
km²	2.59	mi²				
(Mass)	**(Avoirdupois Weight)**			**(Mass)**	**(Troy Weight)**	
grams	0.0648	grains		g	31.103	oz t
g	28.349	oz		g	373.248	lb t
g	453.59	lb		**(Mass)**	**(Apothecaries' Weight)**	
kg	0.45359	lb		g	3.387	dr ap
tonne	0.907	short ton		g	31.103	oz ap
tonne	1.016	long ton		g	373.248	lb ap

Standard Resistance Values

Numbers in **bold** type are ± 10% values. Others are 5% values.

Ohms										Megohms				
1.0	3.6	**12**	43	**150**	510	**1800**	6200	**22000**	75000	0.24	0.62	1.6	4.3	11.0
1.1	**3.9**	13	**47**	160	**560**	2000	**6800**	24000	**82000**	**0.27**	**0.68**	**1.8**	**4.7**	**12.0**
1.2	4.3	**15**	51	**180**	620	**2200**	7500	**27000**	91000	0.30	0.75	2.0	5.1	13.0
1.3	**4.7**	16	**56**	200	**680**	2400	**8200**	30000	**100000**	**0.33**	**0.82**	**2.2**	**5.6**	**15.0**
1.5	5.1	**18**	62	**220**	750	**2700**	9100	**33000**	110000	0.36	0.91	2.4	6.2	16.0
1.6	**5.6**	20	**68**	240	**820**	3000	**10000**	36000	**120000**	**0.39**	**1.0**	**2.7**	**6.8**	**18.0**
1.8	6.2	**22**	75	**270**	910	**3300**	11000	**39000**	130000	0.43	1.1	3.0	7.5	20.0
2.0	**6.8**	24	**82**	300	**1000**	3600	**12000**	43000	**150000**	**0.47**	**1.2**	**3.3**	**8.2**	**22.0**
2.2	7.5	**27**	91	**330**	1100	**3900**	13000	**47000**	160000	0.51	1.3	3.6	9.1	
2.4	**8.2**	30	**100**	360	**1200**	4300	**15000**	51000	**180000**	**0.56**	**1.5**	**3.9**	**10.0**	
2.7	9.1	**33**	110	**390**	1300	**4700**	16000	**56000**	200000					
3.0	**10.0**	36	**120**	430	**1500**	5100	**18000**	62000	**220000**					
3.3	11.0	**39**	130	**470**	1600	**5600**	20000	**68000**						

Resistor Color Code

Color	Sig. Figure	Decimal Multiplier	Tolerance (%)	Color	Sig. Figure	Decimal Multiplier	Tolerance (%)
Black	0	1		Violet	7	10,000,000	
Brown	1	10		Gray	8	100,000,000	
Red	2	100		White	9	1,000,000,000	
Orange	3	1,000		Gold	—	0.1	5
Yellow	4	10,000		Silver	—	0.01	10
Green	5	100,000		No color	—		20
Blue	6	1,000,000					

Schematic Symbols Used in Circuit Diagrams

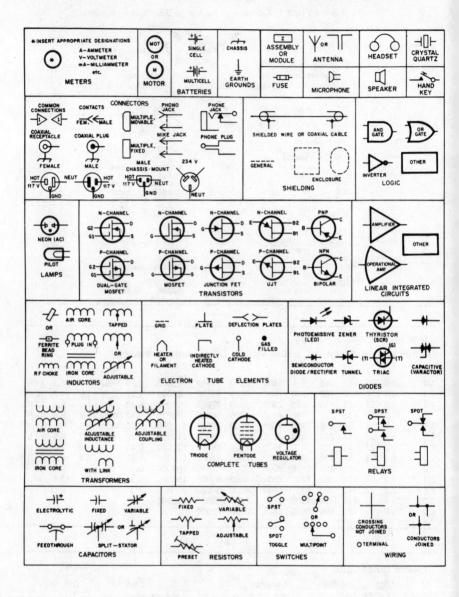

Standard Capacitance Values

pF	pF
0.3	470
5	500
6	510
6.8	560
7.5	600
8	680
10	750
12	800
15	820
18	910
20	1000
22	1000
24	1200
25	1200
27	1300
30	1500
33	1500
39	1600
47	1800
50	2000
51	2200
56	2500
68	2700
75	3000
82	3300
91	3900
100	4000
120	4300
130	4700
150	4700
180	5000
200	5000
220	5600
240	6800
250	7500
270	8200
300	10000
330	10000
350	20000
360	30000
390	40000
400	50000
470	

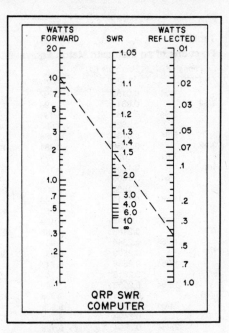

Nomograph of SWR versus forward and reflected power for levels up to 20 watts. Dashed line shows an SWR of 1.5:1 for 10 W forward and 0.4 W reflected.

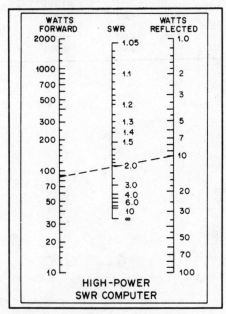

Nomograph of SWR versus forward and reflected power for levels up to 2000 watts. Dashed line shows an SWR of 2:1 for 90 W forward and 10 W reflected.

Table 1

Fractions of an Inch with Metric Equivalents

Fractions Of An Inch		Decimals Of An Inch	Millimeters	Fractions Of An Inch		Decimals Of An Inch	Millimeters
	1/64	0.0156	0.397		33/64	0.5156	13.097
1/32		0.0313	0.794	17/32		0.5313	13.494
	3/64	0.0469	1.191		35/64	0.5469	13.891
		0.0625	1.588	9/16		0.5625	14.288
	5/64	0.0781	1.984		37/64	0.5781	14.684
3/32		0.0938	2.381	19/32		0.5938	15.081
	7/64	0.1094	2.778		39/64	0.6094	15.478
1/8		0.1250	3.175	5/8		0.6250	15.875
	9/64	0.1406	3.572		41/64	0.6406	16.272
5/32		0.1563	3.969	21/32		0.6563	16.669
	11/64	0.1719	4.366		43/64	0.6719	17.066
3/16		0.1875	4.763	11/16		0.6875	17.463
	13/64	0.2031	5.159		45/64	0.7031	17.859
7/32		0.2188	5.556	23/32		0.7188	18.256
	15/64	0.2344	5.953		47/64	0.7344	18.653
1/4		0.2500	6.350	3/4		0.7500	19.050
	17/64	0.2656	6.747		49/64	0.7656	19.447
9/32		0.2813	7.144	25/32		0.7813	19.844
	19/64	0.2969	7.541		51/64	0.7969	20.241
5/16		0.3125	7.938	13/16		0.8125	20.638
	21/64	0.3281	8.334		53/64	0.8281	21.034
11/32		0.3438	8.731	27/32		0.8438	21.431
	23/64	0.3594	9.128		55/64	0.8594	21.828
3/8		0.3750	9.525	7/8		0.8750	22.225
	25/64	0.3906	9.922		57/64	0.8906	22.622
13/32		0.4063	10.319	29/32		0.9063	23.019
	27/64	0.4219	10.716		59/64	0.9219	23.416
7/16		0.4375	11.113	15/16		0.9375	23.813
	29/64	0.4531	11.509		61/64	0.9531	24.209
15/32		0.4688	11.906	31/32		0.9688	24.606
	31/64	0.4844	12.303		63/64	0.9844	25.003
1/2		0.50000	12.700	1		1.0000	25.400

Appendix B

Equations Used in this Book

$$D(mi) = 1.415 \times \sqrt{H(ft)}$$
(Eq. 3-1)

$$D(km) = 4.124 \times \sqrt{H(m)}$$
(Eq. 3-2)

$$Error = f(Hz) \times \frac{counter\ error}{1,000,000}$$
(Eq. 4-1)

$$f_2 = \frac{n_2}{n_1} f_1$$
(Eq. 4-2)

$$1\ volt = \frac{1\ joule}{1\ coulomb}$$
(Eq. 5-1)

$$W = \frac{V^2 C}{2}$$
(Eq. 5-2)

$$W = \frac{I^2 L}{2}$$
(Eq. 5-3)

$$\tan \theta = \frac{side\ opposite}{side\ adjacent}$$
(Eq. 5-4)

$$\sin \theta = \frac{side\ opposite}{hypotenuse}$$
(Eq. 5-5)

$$C^2 = A^2 + B^2$$
(Eq. 5-6)

$$C = \sqrt{A^2 + B^2}$$
(Eq. 5-7)

$$I_R = \frac{E}{R}, \quad I_L = \frac{E}{X_L} \quad and \quad I_C = \frac{E}{X_C}$$
(Eq. 5-8)

$$f_r = \frac{1}{2\pi \sqrt{LC}}$$ (Eq. 5-9)

$$L = \frac{1}{(2\pi f_r)^2 \, C}$$ (Eq. 5-10)

$$C = \frac{1}{(2\pi f_r)^2 \, L}$$ (Eq. 5-11)

$$Q = \frac{X}{R}$$ (Eq. 5-12)

$$Q = \frac{R}{X}$$ (Eq. 5-13)

$$\Delta f = \frac{f_r}{Q}$$ (Eq. 5-14)

$$P = I\,E$$ (Eq. 5-15)

$$P = I^2\,R$$ (Eq. 5-16)

$$P = \frac{E^2}{R}$$ (Eq. 5-17)

$$\text{Power factor} = \frac{P_{REAL}}{P_{APPARENT}}$$ (Eq. 5-18)

$$P_{REAL} = P_{APPARENT} \times \text{Power factor}$$ (Eq. 5-19)

$$\text{Power factor} = \cos \theta$$ (Eq. 5-20)

$$dB = 10 \log \left(\frac{P2}{P1} \right)$$ (Eq. 5-21)

$$R_T = \frac{R1 \times R2}{R1 + R2} \qquad \text{(Eq. 5-22)}$$

$$\beta = \frac{I_c}{I_b} \qquad \text{(Eq. 6-1)}$$

$$\alpha = \frac{I_c}{I_e} \qquad \text{(Eq. 6-2)}$$

$$P = I \times E \qquad \text{(Eq. 7-1)}$$

$$I = \frac{P}{E} \qquad \text{(Eq. 7-2)}$$

$$Z = \frac{E}{I} \qquad \text{(Eq. 7-3)}$$

$$R1 = \frac{E_{dc(min)} - E_Z}{1.1 \, I_{L(max)}} \qquad \text{(Eq. 7-4)}$$

$$P_D = \left[\frac{E_{dc(max)} - E_Z}{R_S} - I_{L(min)} \right] \times E_Z \qquad \text{(Eq. 7-5)}$$

$$R_S = \frac{(V' - V_Z)}{I_{RS}} \qquad \text{(Eq. 7-6)}$$

$$P_{Q1} = I_E \times V_{CE} \qquad \text{(Eq. 7-7)}$$

$$R_{e-b} = \frac{26}{I_e} \qquad \text{(Eq. 7-8)}$$

$$A_V = \frac{R_L}{R_{e-b}} \qquad \text{(Eq. 7-9)}$$

$$R_b = \beta \, R_{e-b} \qquad \text{(Eq. 7-10)}$$

$$A_V = \frac{R_L}{R_E} \qquad \text{(Eq. 7-11)}$$

$$P_{IN} = P_{OUT} + P_D \qquad \text{(Eq. 7-12)}$$

$$\text{Eff.} = \frac{P_{OUT}}{P_{IN}} \times 100\% = \frac{P_{OUT}}{P_{OUT} + P_D} \times 100\% \qquad \text{(Eq. 7-13)}$$

$$R_L = \frac{V_{CC}^2}{2P_O} \qquad \text{(Eq. 7-14)}$$

$$R_L = \frac{V_P}{KI_P} \qquad \text{(Eq. 7-15)}$$

$$F = NF_r \qquad \text{(Eq. 7-16)}$$

$$\text{deviation ratio} = \frac{D_{max}}{M} \qquad \text{(Eq. 8-1)}$$

$$\text{modulation index} = \frac{D_{max}}{m} \qquad \text{(Eq. 8-2)}$$

$$f = \frac{1}{T} \qquad \text{(Eq. 8-3)}$$

$$V_{peak} = V_{RMS} \times \sqrt{2} = V_{RMS} \times 1.414 \qquad \text{(Eq. 8-4)}$$

$$V_{RMS} = \frac{V_{peak}}{\sqrt{2}} = V_{peak} \times 0.707 \qquad \text{(Eq. 8-5)}$$

$$V_{avg} = V_{peak} \times 0.636 \qquad \text{(Eq. 8-6)}$$

$$V_{avg} = V_{RMS} \times 0.899 \qquad \text{(Eq. 8-7)}$$

$$V_{RMS} \times I_{RMS} = P_{avg} \qquad \text{(Eq. 8-8)}$$

$$V_{peak} \times I_{peak} = P_{peak} = 2 \times P_{avg} \qquad \text{(Eq. 8-9)}$$

$$\lambda = \frac{v}{f} \qquad \text{(Eq. 9-1)}$$

$$\lambda(\text{meters}) = \frac{300}{f \text{ (MHz)}} \qquad \text{(Eq. 9-2)}$$

$$\lambda(\text{feet}) = \frac{984}{f \text{ (MHz)}} \qquad \text{(Eq. 9-3)}$$

$$V = \frac{1}{\sqrt{\epsilon}} \qquad \text{(Eq. 9-4)}$$

$$\text{Length (m)} = \frac{300 \text{ V}}{f \text{ (MHz)}} \qquad \text{(Eq. 9-5)}$$

$$\text{Length (ft)} = \frac{984 \text{ V}}{f \text{ (MHz)}} \qquad \text{(Eq. 9-6)}$$

$$\text{Length (feet)} = \frac{492}{f \text{ (MHz)}} \qquad \text{(Eq. 9-7)}$$

$$\text{Length (feet)} = \frac{468}{f \text{ (MHz)}} \qquad \text{(Eq. 9-8)}$$

$$\text{Efficiency} = \frac{(R_R)}{(R_T)} \times 100\% \qquad \text{(Eq. 9-9)}$$

Index

Notes

Notes

Notes

Notes

Notes

Notes

Notes

Notes

Please use this form to give us your comments on this book. Tell us what you liked best about the book, and what improvements you would like to see us make in future editions.

Name

———————————————————————— Call sign ————————

Daytime Phone (   ) ———————————————— Age ————————

Address —————————————————————————————————

City, State, Zip ————————————————————————————

How long have you been licensed? ————————————————————

From ———————————————

———————————————

———————————————

Editor, ACLM
American Radio Relay League
225 Main Street
Newington, CT USA 06111

································· please fold and tape ·······························